The Saber and Scroll Journal

Also from Westphalia Press
westphaliapress.org

The Idea of the Digital University

Dialogue in the Roman-Greco World

The History of Photography

International or Local Ownership?: Security Sector Development in Post-Independent Kosovo

Lankes, His Woodcut Bookplates

Opportunity and Horatio Alger

The Role of Theory in Policy Analysis

The Little Confectioner

Non-Profit Organizations and Disaster

The Idea of Neoliberalism: The Emperor Has Threadbare Contemporary Clothes

Social Satire and the Modern Novel

Ukraine vs. Russia: Revolution, Democracy and War: Selected Articles and Blogs, 2010-2016

James Martineau and Rebuilding Theology

A Strategy for Implementing the Reconciliation Process

Issues in Maritime Cyber Security

A Different Dimension: Reflections on the History of Transpersonal Thought

Iran: Who Is Really In Charge?

Contracting, Logistics, Reverse Logistics: The Project, Program and Portfolio Approach

Unworkable Conservatism: Small Government, Freemarkets, and Impracticality

Springfield: The Novel

Lariats and Lassos

Ongoing Issues in Georgian Policy and Public Administration

Growing Inequality: Bridging Complex Systems, Population Health and Health Disparities

Designing, Adapting, Strategizing in Online Education

Pacific Hurtgen: The American Army in Northern Luzon, 1945

Natural Gas as an Instrument of Russian State Power

New Frontiers in Criminology

Feeding the Global South

Beijing Express: How to Understand New China

The Rise of the Book Plate: An Exemplative of the Art

The Saber and Scroll Journal

Volume 9, Number 1 • Summer (June) 2020

Lew Taylor, editor-in-chief

Westphalia Press
An imprint of Policy Studies Organization

The Saber and Scroll Journal
Volume 9, Number 1 • Summer (June) 2020

Westphalia Press
An imprint of Policy Studies Organization
1527 New Hampshire Ave., NW
Washington, D.C. 20036
info@ipsonet.org

ISBN: 978-1-63391-499-5

Cover and interior design by Jeffrey Barnes
jbarnesbook.design

Daniel Gutierrez-Sandoval, Executive Director
PSO and Westphalia Press

Updated material and comments on this edition
can be found at the Westphalia Press website:
www.westphaliapress.org

TABLE OF CONTENTS

Welcome Letter

Welcome to the Summer 2020 publication of the *Saber and Scroll Journal.* Since the 1960s, gender and women's history has expanded under the pioneering efforts of great female contributors such as Mary Beard (1876–1958), Eleanor Flexner (1908–1995), and Gerda Lerner (1920–2013). Thanks to their efforts, along with many others, women's studies and gender studies have continued to grow and develop into an understanding that women are 50 percent of history. The journal staff hopes you enjoy this season's journal focused on women's history and women historians.

Mary Ritter Beard, an American historian, women's suffrage activist, women's history activist, and author of *Woman as Force of History: A Study in Traditions and Realities* (1946), spent a lifetime along with her husband, historian Charles Beard, researching and writing about women in history. Beard, along with Alice Paul, Eugenia Leonard, and Elizabeth Schlesinger, established the National Women's History Archives. Pioneering women's history, Beard defied the women's suffrage movement that victimized women and pushed for equality with men. Mary thought women should pave their own way and focus on building society and civilizations.

Eleanor Flexner was a respected independent scholar and only one of five identified US History scholars of Women's History in 1970. She is the author of *Century of Struggle: The Women's Rights Movement in the United States* (1959), which combined years of research of the women's suffrage movement, women of color activism, and the women's labor movement. Although Flexner was not associated with the academic community, her impeccable research earned her recognition and respect among scholarly historians, especially women historians, such as Gerda Lerner.

Founder of the first, in American and abroad, Women's Studies MA Program at Sarah Lawrence College in 1972, Gerda Lerner continued to pave the foundation began by Mary Beard and Eleanor Flexner. Lerner entered Columbia University at the age of forty-three in 1963, where she received her MA and PhD in three short years. She chose Columbia due to the bending of rules that allowed her to write her dissertation on women, *The Grimke Sisters*. Upon finishing her degree, she was advised to hide her "special talent," meaning Women's Studies, as she was an excellent social historian. Luckily for the field of women's/gender studies and women historians, Gerda could not push forward with a history where women did not exist.

> It has been my great privilege to be part of the most exciting intellectual movement of the twentieth century.
>
> —Gerda Lerner *Living With History/Making Social Change* (2009)

In this issue, you will find numerous articles about women's history and works by women historians, such as the featured article *Identity in a Teacup: Tea's Influence Over the Lives of British Women in the Nineteenth Century.* Other manuscripts included are *Australian and American Relations in the Southwest Pacific Theater of World War II, The Progression of Women throughout Historiography: A Study of Feminism's Effect on Women as Historians, Not Fit to Breed: Eugenics in Sweden, 1900 to Present, Revisiting the Slaughter-House Cases (1873),* and *The Morality of Genocide: The Holocaust Revisited.* Also in this issue, you will find some excellent book and museum reviews.

If you are interested in submitting an article, book review, or museum review for consideration, please check out the submission guidelines included in this issue. Do you have any comments to share? If so, please submit those as a "Letter to the Editor."

Gina Pittington, AMU Alumni

Women's/Gender Studies and Holocaust/Genocide History

Dear Readers,

Beginning this September, we will feature a new section in the journal: "Letters to the Editor." In this section, we will feature letters from you, our readers. We want to know what you think! Send us your thoughts, comments, kudos, and suggestions. By understanding what you like and your suggestions for improvement, we hope to give our authors feedback and also to continue to improve the journal.

We are not looking at any set length, but we do ask that all feedback be constructive. We will also answer questions you might have. We may not have space to publish every comment, but will give you feedback, whether or not we are able to print your submission.

Please send your submission to "Letters to the Editor" c/o:
EICSaberandScroll@apus.edu

Once again, we'd love to hear what you think!

Chris Schloemer
Senior Editor, *Saber Scroll Historical Journal*

Slavery in New York: Through the Lens of James Fenimore Cooper's Written Works

Sharon Powell
Herkimer Home State Historic Site

Abstract

In this article, I propose to analyze James Fenimore Cooper's written works, including *The Wept of Wish-ton-Wish* and "A Defense of Slave-Owning America," in terms of the ways that these works, and others, represent Cooper's attitude and the attitude of Americans toward slavery and the black community during the late eighteenth and nineteenth centuries, particularly those residing in the state of New York.

Keywords: slavery, Palatine Germans, Herkimers, James Fenimore Cooper, abolitionism, amalgamation, mixed race, democratic nation

La esclavitud en Nueva York: a través de la lente de las obras escritas de James Fenimore Cooper

Resumen

En "La esclavitud en el estado de Nueva York: a través de la lente de las obras escritas de James Fenimore Cooper", propongo analizar las obras escritas de James Fenimore Cooper, incluyendo The Wept of Wish-ton-Wish y "A Defense of Slave-Owning America" en términos de las formas en que estas obras, y otras, representan la actitud de Fenimore hacia la esclavitud y la comunidad negra, así como la actitud de los estadounidenses a fines del siglo XVIII y XIX, particularmente aquellos que residen en el estado de Nueva York.

Palabras clave: esclavitud, alemanes palatinos, Herkimers, James Fenimore Cooper, abolicionismo, amalgama, raza mixta, nación democrática

纽约州奴隶制：透过詹姆斯·费尼莫尔·库珀作品视角

摘要

在《纽约州奴隶制：透过詹姆斯·费尼莫尔·库珀作品视角》一文中，我提出分析詹姆斯·费尼莫尔·库珀的作品，包括*The Wept of Wish-ton-Wish* 和"支配奴隶的美国防御"，分析这些作品及其他作品以何种方式代表费尼莫尔对奴隶制和黑人社群所持的态度，以及18世纪末和19世纪时期美国人的态度，尤其是那些居住在纽约州的美国人。

关键词：奴隶制，Palatine Germans，Herkimers，詹姆斯·费尼莫尔·库珀，废奴主义，异族通婚，种族混合，民主国家

Slavery has taken many forms in different parts of the world throughout history. In 1619, the first African arrived in the colonies—in Virginia, specifically. The spread of slavery—particularly the race-based slavery that became a divisive institution in the nineteenth century—was not inevitable. Early American colonists used various forms of free labor, including the subjugation of Native Americans and the prolific use of indentured servitude. In the central New York region, the setting of much of Cooper's writings, there was an influx of indentured servants during the early eighteenth century. Palatine Germans fled their homes in the southern Palatinate region of Germany for a variety of reasons: floods, famine, religious persecution, the ongoing wars of Louis XIV of France, and the circulation of pamphlets promoting opportunities in North America. Unlike black slaves, who were introduced to the colonies forcefully, German refugees were presented with opportunities of freedom and land in exchange for labor. Palatine refugees were sent to New York by Queen Anne, who negotiated their terms of servitude with then-Governor Hunter; the Palatines were to live in the Hudson Valley on the land of Robert Livingston to produce tar pitch out of local pine trees for the Royal British Navy. As a result of this unique, collective experience, Palatine refugees—some of whom later became slave owners—viewed slavery differently than much of the American public.

James Fenimore Cooper believed in the principles of abolitionism. However, Cooper did not view slavery as a national crime, as many hardline abolitionists did. Cooper thought that slavery would die a natural death in due course.[1] Thus, Cooper was content to explain the moral and ethical evils of the institution of slavery, while predicting that its demise was imminent.[2]

In Cooper' *Defense of Slave-Owning America*—in which Cooper defended the uniquely American institution from foreign criticism— Cooper argued that slavery was viewed by the majority of Americans as an evil, immoral practice, but that it remedied more easily in theory than in actual practice. Cooper's argument was a reflection of the notion shared by many Americans that slavery is not a palatable practice but that blacks are also not equal in any fundamental way. This sentiment also reflected that of the Founding Fathers at the inception of the American nation. The Founding Fathers also promoted a conflicted sentiment: slavery was considered immoral and detrimental to the principles of a democratic government, but blacks were considered by the majority to be innately inferior to whites Thomas Jefferson argued this very idea Jefferson also bought and sold slaves advertised for fugitives, and ordered the punishment of fugitive slaves.[3]

James Fenimore Cooper, National Portrait Gallery, Smithsonian Institution

Cooper also—although he did not approve of slavery—chose to defend slave-owning America against foreign criticism, as he argued that others did not understand the circumstances of its existence in an otherwise civilized and enlightened nation.[4] There was an abundance of available land to cultivate and develop and not enough able bodies to work it. Thus, slavery and indentured labor became a fundamental aspect of the developing nation. As such, there existed within colonial America and the American nation after the Revolutionary War a dual view and relationship with slavery.

In the Mohawk Valley, there was a unique perspective of slavery. The Palatine Germans, who eventually settled in the Mohawk Valley, the Schoharie Valley, and parts of Pennsylvania, were formerly indentured servants of the British government. Various Palatines amassed wealth as a result of land grants and trade. With wealth came the necessity for labor, and the Palatines used indentured labor, tenant farmworkers, and the enslaved. The Herkimers—a local prominent Palatine family—amassed such wealth that they acquired a reported thirty-three slaves, which was an unusually large number in this region. The will of General Nicholas Herkimer, who left upon his death the bulk of his estate to his brother George,

Herkimer Home State Historic Site, Little Falls, NY

included the stipulation that "my negro slaves, Dick, Sam, and Mary to be well used in their ages in remembrance of good and faithful service—should they be misused such to be taken from my brother George by my executors."[5] As a result of the experience of the Palatines in a position of servitude, there was a different view of slavery in which there was a willingness to free enslaved persons without the fear of violent retaliation, especially if their treatment violated what was considered fair. This sentiment was not widely shared; Cooper expressed his distaste for slavery; but shared the concerns of many of his contemporaries regarding the threat of slave rebellions or violence perpetrated by freed slaves.

Until the eighteenth century, the image of Africans was generally positive. They were farmers and cattle-breeders; they had industries, arts, and crafts, governments, and commerce. In addition, Africans had immunity to Old World diseases. They were better laborers, and they had nowhere to escape to once transplanted to the New World. The colonists themselves came to believe that they could not survive without Africans.[6] During the infancy of the American Republic, there were differing views of the practice of enslavement. Certainly, slavery was inconsistent with the values espoused in the Declaration of Independence: "We hold these truths to be self-evident, that all men are created equal, that they are endowed by their Creator certain inalienable rights, that among these are life, liberty, and the pursuit of happiness." This declaration was in direct contrast with the practice and institution of African enslavement. James Madison asserted that slavery was indeed inconsistent with Republican ideals.[7] Madison was not alone in his abhorrence for slavery; Benjamin Franklin, Samuel Adams, John Adams,

John Jay, Thomas Jefferson, and George Washington were among the multitude of founders and statesmen who indicated a distaste—if not outright aversion—to slavery.[8] These men were unable to translate this aversion to any political or social action against slavery, although there were a number of weak attempts made to initiate emancipation legislation. For example, in 1789, Benjamin Franklin composed and published numerous essays in support of abolishing slavery. Franklin's final public act was to send a petition to Congress on behalf of the Pennsylvania Society for Promoting the Abolition of Slavery, which called for the for the abolition of slavery and an end to the slave trade. The petition, which was signed on February 3, 1790, asked the first Congress to "devise means for removing the Inconsistency from the Character of the American People" and to "promote mercy and justice toward this distressed Race."[9]

General public perception of the black community in the eighteenth century was negative and unfavorable. Blacks were viewed, almost consistently, as intellectually and morally inferior, lustful, lazy, and prone to violent tendencies.[10] Concerning their physical inferiority, blacks were described by Thomas Jefferson as "aesthetically inferior ... they can't blush, and they themselves prefer white features ... as uniformly as ... the preference of the Orangootan for the black women over those of his own species."[11] This general characterization echoes the sentiment of Jefferson's fellow citizens. Blacks were viewed as innately intellectually, physically, and morally inferior.

Cooper's portrayal of enslaved and free blacks alike reflects the confusing nature of America's relationship and view of both slavery and the notion of universal equality. Although the practice of enslavement was considered morally deficient, the idea of racial equality was not accepted by the majority of Americans in the eighteenth century; thus, there was a disconnect between the need to eradicate slavery for the sake of the nation's morality and the need to maintain a racial hierarchy. Despite this, it was assumed that slavery would not exist in perpetuity, as it was an immoral practice. As such, it was imminent that slavery would be abolished in some manner. Jefferson, who was himself a slave owner, spoke in favor of abolition. There was a widespread notion of emancipation, with deportation soon thereafter. In 1813, a freeman in Massachusetts proposed—through the Senator of Massachusetts—that he should be permitted to take free blacks with him to the British colony of Sierra Leone. When this proposal went before the House, there was a consensus that the measure would rid the country of "free blacks, a part of our population which we could well spare." Delegations that expressed this sentiment came from the colonies of New Hampshire, New York, Rhode Island, Connecticut, Delaware, Vermont, New Jersey, and Massachusetts. This measure was ultimately defeated, but not because there was no consensus, but instead because there was a disagreement regarding embargo policies. In addition, the notion that blacks—slaves, in particular—were prone to violence was widespread and used to justify subjugation.

The term "miscegenation" did not exist in the 1820s when Cooper was writing some of the examined novels. This term was invented in the 1860s, around the same time that the discussion about modern racism was initiated, with an emphasis on "genetics" and "eugenics." The primary goal was to compensate for the effects of the pending Emancipation Proclamation.[12] Before 1863, the prominent terms for describing interracial children were mongrelization and amalgamation. Another phrase that Cooper frequently used was "intermingling of blood," which is less derogatory and more neutral.[13]

Cooper had a more accepting attitude toward the mixing of races concerning whites and Native Americans. In the *Wish of Wept-Ton-Wish,* Cooper emphasized the similarities between the settlers and the Narragansetts. This narrative seems to suggest that the similarities between the two races may prove to be the basis for intermarriage and intermingling. Metacom—in *Wish*—calls Naramattah—the white captive who has married Conanchet—"One who is neither white nor red."[14] However, Conanchet later forsakes his wife and child, and spoke about the necessity of keeping the races separated in a poignant speech. Conanchet compared different species of trees and argued that mixing two completely different types of trees that were not meant to grow together had angered the Great Spirit. Cooper was not only opposed to intermarriage outside of his texts but he also emphasized this concept within the boundaries of his fictional works through his characters' interactions with different races and the resulting consequences of mixed-race interactions.

James Fenimore Cooper did not approve of slavery. However, he did not engage in a direct attack on the practice itself, especially as it was a decidedly American practice during the nineteenth century. This philosophy was similar to the approach that ordinary Americans took with regard to slavery. Slavery was a practice that was condemned easily and sentenced to a short life of unfortunate necessity. However, outside criticism was unwelcome, as slavery was a pervasive practice within both the North and the South in the seventeenth and eighteenth centuries.

Cooper was offended by critiques of the United States and the continued practice of slavery, which was of a contradictory nature when viewed through an international lens. Cooper also defended the treatment of slaves in the United States, arguing that, "Doubtless there are many abuses, but in general they are at least well clothed and lodged, and far better fed than half the peasants in Europe."[15] Cooper argued multiple times that the American slave was far better off than the European peasant, in both treatment and conditions of labor. Cooper also argued that if not for the actions of the European colonial powers, slavery would never have been initiated in North America, nor would it have been needed. Further, one of the grievances expressed as a reason for the Declaration of Independence from England was the repugnant nature of slavery; during the Revolutionary period, Cooper argued, slavery saturated all of the colonies; after declaring independence, fewer areas of the country had slavery as a common practice.[16] Cooper maintained throughout this text that the black slave and the free black in America had more rights than some European citizens, stating "They are provided for in their age, are never seen cumbering the approach to the alters, objects of misery and disgust, imploring alms and exhibiting their ails and wants."[17] Thus, there were numerous justifications for slavery, although Cooper did not believe that slavery could—or should—continue to exist within a free, democratic society.

Cooper worked black characters into the main story plots in his novels, which was a progressive action on his part as a novelist. However, Cooper also created a prescribed, caricatured format for his black characters: misshapen, dense, simple, superstitious, and always smiling—even in the face of danger or serious situations.[18] This is also found in Coopers *Defense of Slave-Owning America*, where it is stated that blacks are a "race proverbial for their light-heartedness. The laugh of the negro is merriment itself."[19] This is intended to demonstrate that the position of a slave—or a black society member—was not a cause for misery. It was the institution itself that was unjust. This depiction of blacks was present in various other works of Cooper. In the novel, *The Red Rover,* the character of Cassandra is a minor character; she is a slave, but she also plays an important role. She succeeds in helping rescue her mistress, Gertrude Grayson, through an act of rebellion, by distracting the men aboard the pirate ship, against the objections of Wilder. In this manner, Cooper portrays Cassandra as a rebellious slave, which aligns with the fear that many Americans had regarding the enslaved: that they would rise up in rebellion against their respective authority figures, either individually or en masse.[20] This fear is a vivid example of the contradictory nature of the public assessment of slavery. On the one hand, slavery is immoral and contradicts the very principles of individualism, freedom, and democracy. However, those who have been enslaved are viewed as prone to violence and rebellion as a reaction to their unfortunate condition, so there must be a remedy before slavery can be abolished, as there are safety concerns for the remainder of society.

Cooper also allowed for some characters to display an open hostility toward white characters. In addition, there are instances where white characters defend the "good qualities" of black characters, bemoaning the fact that they are black and thus innately inferior despite their good qualities.[21] Cooper placed all black characters—whether free, slaves, or of mixed race—in subordinate roles throughout the texts, although some actions taken by these characters contribute to the success of the main white characters. This is a reflection of the contradictory sentiment regarding the position of blacks in society, social equality, and the injustice and immorality of slavery. These sentiments were in contrast with one another, but continued to be prevalent.

James Fenimore Cooper was an important figure in central New York.

Cooper's texts highlighted the surrounding valleys and the rich history in this part of the country. These texts also mirrored the thoughts of many Americans regarding slavery—that it was an unethical practice that would eventually die a natural death—and the innate inferiority of all blacks, both free and enslaved. Cooper also echoed the idea that was shared by many regarding the mixing of the races and used the medium of fictional prose to espouse the dangers of intermarriage. Cooper defended the American practice of slavery by solidly placing the blame for the practice on European colonial powers and highlighting the concept of the diminished nature of American slavery after the nation declared independence from England. Simultaneously, Cooper defended the treatment of slaves—living within an immoral institution that he despised—by characterizing blacks as naturally suited to that position, as they were biologically programmed to be of a positive, happy nature under a multitude of circumstances. There are many contradictions to the position that Cooper takes concerning slavery and the treatment of blacks under the slave system; these contractions are representative of the broad—and localized—public sentiment.

Bibliography

Becker, George J. "James Fenimore Cooper and American Democracy." *College English* 17, no. 6 (March 1956): 325–34.

Dorsey, Peter A. "To "Corroborate Our Own Claims": Public Positioning and the Slavery Metaphor in Revolutionary America." American Quarterly 55, no. 3 (2003): 353-86.

Falk, Cynthia G. "Forts, Rum, Slaves, and the Herkimers' Rise to Power in the Mohawk Valley." *New York History* 83, no. 3 (Summer 2008): 221–34.

"Fenimore Cooper's Defense of Slave-Owning America." *The American Historical Review* 35, no. 3 (April 1930).

Kates, Jr. Don B. "Abolition, Deportation, Integration: Attitudes Toward Slavery in the Early Republic." *The Journal of Negro History* 53, no. 1 (January 1968): 33–47.

Mautner Wasserman, Renata R. "Literature and Cultural Identity in the United States and Brazil, 1830–1930." In *Exotic Nations, Chapter Title: James Fenimore Cooper and the Image of America.* Cornell University Press, 1994.

O'Daniel, Therman B. "Cooper's Treatment of the Negro." *Phylon (1940-1956)* 8, no. 2 (2nd Quarter, 1947).

Olson, Edwin. "Social Aspects of Slave Life in New York." *The Journal of Negro History* 26, no. 1 (January 1941): 66–77.

Smedley, A. *Race in North America: Origin and Evolution of a Worldview.* Boulder: Westview Press, 1993.

———. "Origin of the Idea of Race." *Anthropology Newsletter,* November

1997. https://www.pbs.org/race/000_About/002_04-background-02-09.htm.

Wallace, James D. "Race and Captivity in Cooper's The Wept of Wish-Ton-Wish." *American Literary History* 7, no. 2 (Summer 1995).

Watkins, William H. "Dean Miller of Little Falls, New York: A Mohawk Valley Slave." *Journal of the Afro-American Historical and Genealogical Society.*

Notes

1 "Fenimore Cooper's Defense of Slave-Owning America," *The American Historical Review* 35, no. 3 (April 1930): 575.

2 Ibid.

3 William W. Freehling, "The Founding Fathers and Slavery," *The American Historical Review* 77, no. 1 (February 1972): 82.

4 "Fenimore Cooper's Defense of Slave Owning America," 576.

5 William H. Watkins, "Dean Miller of Little Falls, New York: A Mohawk Valley Slave." *Journal of the Afro-American Historical and Genealogical Society.*

6 Audrey Smedley, "Origin of the Idea of Race." *Anthropology Newsletter,* November 1997. https://www.pbs.org/race/000_About/002_04-background-02-09.htm.

7 Don B. Kates, Jr., "Abolition, Deportation, Integration: Attitudes Toward Slavery in the Early Republic." *The Journal of Negro History* 53, no. 1 (January 1968): 33–47.

8 Ibid., 34.

9 Petition from the Pennsylvania Society for Promoting the Abolition of Slavery to Vice President John Adams signed by Benjamin Franklin (front), February 3, 1790; Records of the US Senate, RG 46.

10 Kates, "Abolition," 40.

11 Ibid., 40.

12 Ibid., 191.

13 Ibid., 192.

14 Cooper, 204.

15 "Fenimore Cooper's Defense of Slave Owning America," 579.

16 Ibid., 578.

17 Ibid., 579.

18 Kates, 39.

19 "Fenimore Cooper's Defense of Slave Owning America," 579.

20 Therman B. O'Daniel, "Cooper's Treatment of the Negro," *Phylon (1940-1956)* 8, no. 2 (2nd Quarter, 1947): 166.

21 Ibid., 175.

Identity in a Teacup: Tea's Influence Over the Lives of British Women in the Nineteenth Century

Caitlin Khan
American Military University

Abstract

This paper studies the link between the development of the social aspect of the tea table and the roles it allowed upper-class English women to play within the set framework of nineteenth century British society. Literature of the period clearly indicates that the tea table provided wealthy English women with an identity in society. Worsening political relations between China and Britain restricted the availability of Chinese tea, which led to the development and marketing of Indian tea. Advertising for newly developed Indian teas, which primarily targeted housewives and upper-class women, demonstrates the significance of the authority that English women held over home purchases. In studying the culture surrounding tea as both a social activity and political message, there comes a greater understanding of women and their positions in nineteenth century British upper-class society.

Keywords: tea, upper-class women, social etiquette, Britain, India, China, literature, imperialism, temperance

Identidad en una taza de té: la influencia del té sobre la vida de las mujeres británicas en el siglo XIX

Resumen

Este artículo estudia el vínculo entre el desarrollo del aspecto social de la mesa del té y los roles que permitió que las mujeres inglesas de clase alta desempeñaran dentro del marco establecido de la sociedad británica del siglo XIX. La literatura del período indica claramente la idea de que la mesa de té proporcionaba a las mujeres inglesas ricas una identidad en la sociedad. El empeoramiento de las relaciones políticas entre China y Gran Bretaña restringió la disponibilidad de té chino, lo que condujo al desarrollo y comercialización del té indio. La publicidad de los tés indios recientemente desarrollados que se dirigieron principalmente a las amas de casa y las mujeres de clase alta demuestra la importancia de la autoridad que las mujeres inglesas tienen sobre las compras de viviendas. Al estudiar la cultura que rodea al té como actividad social y mensaje político, se llega a una mayor comprensión de las mujeres y sus posiciones en la sociedad de clase alta británica del siglo XIX.

Palabras clave: té, mujeres de clase alta, etiqueta social, Gran Bretaña, India, China, literatura, imperialismo, templanza

茶杯中的身份认同：19世纪茶对英国女性生活产生的影响

摘要

本文研究了19世纪英国社会背景框架下茶桌在社交方面的发展与上流阶层英国女性因茶桌而发挥的作用之间的联系。该时期的文献清晰显示，茶桌为富裕的英国女性在社会中提供了一种身份。中英之间恶化的政治关系限制了中国茶的供应，导致印度茶的发展与推广。为新研制的印度茶（以家庭主妇和上流阶层女性为目标群体）进行推广，证明了英国女性在家庭购买方面所拥有的支配权的重要性。通过研究茶在作为一项社交活动与（传播）政治信息方面的文化意义，便更能理解19世纪英国上流阶层社会的女性及其地位。

关键词：茶，上流阶层女性，社交礼仪，英国，印度，中国，文学，帝国主义，禁酒

For over two hundred years, the humble cup of tea symbolized British high society. Tea's exotic origins disappeared due to clever advertising campaigns that promoted the beverage as uniquely British. Trade imports followed the growing popularity of tea to encompass the expanding British Empire. The popularity of tea among the upper classes encouraged members of the middle class to adopt the drinking habit as well. By the nineteenth century, the commodity could be found in English houses from all classes. While tea remained popular across all levels of society and among both genders, it gained a unique fascination and appreciation from British upper-class women. Seen through the literature of the era, tea emerges as a figure in its own right, as British authors used tea as a device to demonstrate the comforts of home, courtship rituals, and the strength and size of the British Empire. As tea culture and etiquette developed over time, upper-class women used the opportunities provided to use the tea table advantageously. Although it could have been nothing more than an addition to the household menu, tea instead became the platform for women to define their social standing, cultivate their moral reputations, and demonstrate their imperial patriotism.

From its first arrival in England, tea became primarily associated with women. In a move to rewrite history, tea's introduction to the English court is often attributed to Catherine de Braganza, following her marriage to King Charles II of England in 1662. The earliest known mention of tea in literature is found in the poem, "Of Tea, Commended by Her Majesty," written by Edmund Waller in honor of Queen Catherine's birthday in 1690.[1] Waller credits the queen with blessing his nation with the beverage.[2] The connection to the queen is particularly symbolic; her inability to produce a royal heir can be viewed as less tragic when considering her gift of tea to the English people. While the queen certainly popularized tea as a beverage, allowing its consumption and the demand for it to rise, tea arrived in England several years before she did. London newspapers first advertised tea in the September 1658 edition of *The Gazette*, as a product from the East India Company.[3] Victorian Era historian Agnes Strickland (1796–1874) wrote, "The queen brought tea with her to England—always used tea—and tea in the course of a century became the national drink of the ladies of England," acknowledging Catherine's influence over the royal court's adoption of tea, rather than her introduction of it to the country at large.[4] By the period in which Strickland wrote, tea completely permeated everyday culture for women across all levels of society. Strickland correctly recognized tea's importance with the women of British society through the enjoyment it created in those who drank it and the culture that came out of the beverage.

At first, tea could only be enjoyed by the very wealthy, since limited supply and difficulties in procuring shipments of tea contributed to its high cost.[5] The lengthy sea voyage and China's trade policies only allowed for small shipments to arrive in England. Consumer demand commenced slowly as well; in 1669, the East India Company ordered only one hundred pounds of tea for sale in England.[6] By 1703, however, the demand of the British upper class grew to 75,000 pounds of tea and within ten years that number nearly doubled.[7] The primary demand for tea came from Englishwomen of means. Due to the costly nature of tea, women considered it a luxury item. Women stored tea in their own bedrooms or private sitting rooms, known as closets, where it was not only protected, but also served and enjoyed.[8] As the practice of drinking tea in sitting rooms became more commonplace in households of the upper class, it led to tea's future association with small, intimate gatherings between women of influence.[9] In the 1706 play, *The Way of the World,* William Congreve shares this view of tea through the female heroine who proclaims, "to have my Closet ... to be sole [Empress] of my Tea-Table."[10] The title Empress both represents the wealth and luxury of drinking tea and references the exotic origins of the beverage. Tea symbolized the ultimate extravagance, allowing women to temporarily act the part of an empress, even one over so small a territory as a tea table in their closet.

Increasing demand for the extravagance of tea brought the East India Company to seek out more luxury goods to be sold to the British upper class. As the East India Company expanded trade profitability, a large number of exotic items found increasingly widespread use among British households.[11] The growing wealth and influence of the company, driven by the ever-multiplying demand for tea, brought the item from its position as a limited luxury commodity to an indulgence that few in upper-class society could live without.[12] Advertisements that promoted tea as a medicinal herb increased interest in and demand for the beverage and various tea-related paraphernalia. The East India Company imported porcelain tea sets from China and Japan, which were highly sought after by wealthy English women. Poor packing and shipping methods left many porcelain items damaged, however, which only raised the cost of these goods.[13] Another commodity enhanced by the popularity of tea was sugar, which made the sugar plantations in the British West Indies significantly more profitable.[14] British citizens each consumed an estimated twenty pounds of sugar in the 1790s in a variety of foods; however, sweetened tea remained the largest contributor to this number.[15] Each of these imported goods was taxed by the British government, bringing in more money to their economy.

Originally, coffee houses and apothecaries sold imported gallons of pre-made tea, making taxation on the commodity difficult to determine:

> in 1670-1 a duty of eighteenpence was imposed upon "every gallon of Tea made and sold, to " be paid by the makers thereof." This tax is a curious illustration of the manner in which Tea was originally sold. It was, however, soon found that a tax of this nature was very troublesome to collect, and open to any amount of evasion; so in 1688 the excise upon the liquor was repealed and heavy Customs duties imposed upon the imported Teas instead.[16]

Tea's imported value between 1789 and 1793 reached almost £12,000,000, earning the British government 6 percent of the entire national budget from import taxes alone.[17] Previously, the largest import earning came from calico fabric, which only sold around £3,000,000 annually.[18] Despite the cost of the commodity and its taxes, by 1816, the British had completely integrated tea as a common household item among the upper and middle classes; the novel *Pride and Prejudice* frequently and nonchalantly references tea and the tea table, indicating that the practice was so common that it did not warrant detailed description.[19]

The rise of the upper middle class in the eighteenth century generated a cultural shift for men and women.[20] The nouveau riche broke the understood distinction between levels of society; women who originally functioned as assistants to family business enterprises suddenly found themselves thrust into a life of luxury with a new set of rules.[21] As luxury items like tea became more of an everyday commodity, societal standards evolved. Upper-class society dictated behavioral expectations for wealthy women and men. Wealthy and titled women therefore existed as charming and accomplished ladies of leisure. Likewise, wealthy and titled men were expected to be gentlemen whose only occupation involved estate management and politics.[22] Despite their newfound wealth, men and women of the middle class remained outside of the upper class, having come from a trade background.[23] This new class of English men and women used their positions of wealth to imitate those in the upper classes by living lives of leisure and gentility. Austen references this new class in *Pride and Prejudice* through the character of Sir William Lucas, who "formerly in trade in Meryton, where he had made a tolerable fortune, and risen to the honour of knighthood."[24] To accompany his new position in society, Sir Lucas left his trade to allow his family to live in leisure.[25] The cultural standard of leisure as a way of life brought many women to adopt social causes.[26]

Tea gatherings existed as an acceptable outlet for socialization between wealthy middle-class women who lived firmly on the fringes of high society. Copying society's examples in their new positions of wealth and leisure, tea became more of an event rather than a mere beverage. William Thackeray wrote of the role tea played in the lives of women,

> What a part of confidante has that poor teapot played ever since the kindly plant was introduced among us! What myriads of women have cried over it, to be sure! What sickbeds it has smoked by! What fevered lips have received refreshment from it! Nature meant very kindly by women when she made the tea plant; and with a little thought, what a series of pictures and groups the fancy may conjure up and assemble round the teapot and cup.[27]

Thackeray recognized the significance that tea occupied in the daily existence of women during his time. Although the plant in nature came from humble origins, it served a greater purpose as seen through the lives of women. Assembling around the tea table gave women not only a social outlet, but also provided a role in society. Some women used this role to exert limited influence over business by inviting tradesmen to tea.[28] Georgian era gentlewoman Elizabeth Shackleton frequently welcomed tradespeople to her home to discuss estate business, although she maintained a position of condescension due to her social superiority.[29]

By the nineteenth century, tea drinking had been established as expressly British in nature, despite the fact that no tea crops existed on English soil. Even in 1870, the licensed Victualler's Tea Association could not entirely explain the speed with which English homes adopted tea as their beverage of choice: "The late rise and present magnitude of the British Tea trade are among the most extraordinary phenomena in the history of commerce."[30] Early in tea sales, the beverage could be purchased from practitioners of medicine, which only confirmed tea advertisements extolling the healthy benefits of the beverage.[31] Seventeenth century poet Waller recognized the benefits of tea drinking, "The Muse's friend, tea does our fancy aid, /Repress those vapors which the head invade, /And keep the palace of the soul serene."[32] Through these lines, it is evident that Waller appreciated the benefits that tea provided in clearing his head, but also the soothing effect it had on his mood. Although caffeine was not yet discovered, the energy-boosting quality of tea encouraged sales over alcoholic beverages.[33] Jane Austen wrote about tea's energy-inducing qualities in letters to her sister, suggesting that an acquaintance with sleeping issues might benefit from fewer cups of tea.[34] Within the upper classes, tea drinking provided a much-needed boost to energy after late night society events.

Less than two hundred years after its introduction, tea drinking had become commonplace in British society.[35] Because society already firmly ingrained tea into their social lives, women were able to use the culture surrounding the drinking of it to participate in politics despite not yet having the right to vote. Women's social activism, through the temperance movement, adapted to include more political statements in support of British imperialism. One such statement came out of the tea industry itself. China's strict trading policies hampered the availability of tea in England, causing women to boycott Chinese tea and products in favor of more British commodities. A successful advertising campaign in England during the Second Opium War (1856–1860) encouraged British superiority compared to the "barbaric Chinese people."[36] This perception of Chinese barbarism only increased after political reports were released to inform the English public of instances in which Hong Kong natives attempted to poison foreigners through adulterated bread.[37] Fears that imported tea would soon follow led women from all classes across Britain to reject Chinese tea in favor of newly developed Indian varieties.[38]

Tea found within the Assam province encouraged the British to colonize the area, despite a previous lack of interest in doing so prior to the published discovery of tea in 1839.[39] Queen Victoria publicly proclaimed her preference for Assam tea as a superior product to Chinese tea in 1839, which spurred more women across England to seek out Indian teas.[40] In order to meet England's ever-increasing thirst for tea, the East India Company planted tea throughout regions of India. This fortuitously coincided with souring relations between Britain and China.[41] Governor

General of India William Bentinck established tea plantations within East India controlled territory in order to counteract the Chinese monopoly on the product.[42] However, the first East India Company-grown teas failed to catch public interest, as they lacked the depth of flavor Chinese tea exhibited.[43] The Indian tea trade uniquely focused its attention on women, paying them to spread word of the tea to their friends and family in order to promote the product.[44] Tea merchants and planters hired female tea tasters to develop tea blends that appealed to the feminine public.[45] Advertisements in newspapers and ladies magazines boasted Indian tea as British products, as opposed to the foreign product of China.[46] These advertisements focused their attention on women by portraying cozy domestic scenes on boxes of tea that indicated a connection between home life and the state of the British Empire.[47] By purchasing Indian tea, British women demonstrated their patriotism. This campaign proved so successful that Indian-grown teas dominated the British home: "the Britisher prefers the Empire-grown teas of India and Ceylon," revealing the power of women's influence.[48] Despite the success of the Indian tea campaign, Chinese tea sales persisted, albeit on a diminishing scale: between 1867 and 1907, imports of Chinese tea steadily declined by nearly one hundred million pounds.[49]

Outside of the political realm, teatime served differing rituals and functions and was enjoyed across society according to the culture and rituals of each class. The development of ritual surrounding the upper-class tea table quickly replaced the beverage as the status symbol of the day. While tea, the beverage, still played an integral role, it became the ceremonies and rituals surrounding the beverage that demonstrated a wealthy individual's social status and respectability. As tea became less of a limited and exclusive product, the service and ceremony surrounding it grew to incorporate the symbols of wealth tea once accomplished on its own. Upper-class women used tea culture as a symbol of their wealth, displaying intricately decorated tea caddies and silver or fine porcelain tea services, "from the solid silver urn, of antique pattern, and the massive pot of the same metal, to the thin porcelain cups, dark with purple and gilding."[50] English pottery manufacturers developed these beautifully described new teacups complete with handles and saucers, rather than continuing to use cups from China, which were more akin to bowls.[51] This new service ware disassociated tea with Asia in the mind of the British public, allowing Britain to claim tea as their own.[52] To display their new tea sets, upper- and middle-class women sent out formal invitations to friends and acquaintances for tea. Whereas the seventeenth century tea gatherings took place in a woman's closet, two hundred years later, these parties belonged in a large formal parlor.[53]

The lower classes also enjoyed drinking tea, although it existed primarily as a simple beverage to accompany a meal without any of the ritual and pomp that accompanied tea drinking by wealthier British citizens.[54] Supplemented with sugar, tea served as fuel to maintain men and women in their hard labor.[55] They also added milk or cream to their tea to stretch the nourishment and contents of the cups, a practice initially started by wealthy families at the breakfast table.[56] In writing a letter to her sister, Jane Austen commended a new acquaintance, "there are two traits in her character which are pleasing,—namely, she admires Camilla, and drinks no cream in her tea,"[57] establishing that not all additions to the teacup found appreciation among society. Although Austen did not socialize in aristocratic circles, her commentary on cream's addition to tea demonstrates not only a distinction between classes, but also the influence of the upper class upon the middle class. The tea table provided women with a role in society, as demonstrated by Fanny Price in *Mansfield Park*.[58] A poor relation taken in by Lady Bertram, Fanny serves as companion to her aunt, where part of her duties includes making and serving tea to the family.[59] In contrast, the poor Price home served tea solely as an accompaniment to the evening meal amid family chaos and without ceremony.[60] To the poor, tea existed merely as a beverage.

Far removed from the functionality of tea served for the working class, formal tea parties provided women with the opportunity to demonstrate their abilities as hostess. While the purpose of gathering for tea fundamentally remained the same, for women to socialize among their peers, the circumstances surrounding the time became more of an event. Fashion developed new gowns designated for teatime, presenting upper-class women with the opportunity to look as elegant as their surroundings while they sipped tea from beautiful and delicate cups.[61] Women dominated these elegant and sophisticated gatherings, displaying the good breeding and etiquette trained into them since infancy.[62] Upper-class society women used the developing tea culture as an opportunity to express their social status to friends of their own class. Beautifully expensive tea services gave the women of upper-class society the chance to interact with others of similar standing and to display their wealth. Tea gowns were specifically designed to display both the wealth and beauty of their wearers during these tea events.[63] The purpose for these gatherings served not only to associate with other women of their peer group, but also to allow women to socialize with eligible men.[64] After enjoying refreshments, young women demonstrated their musical accomplishments for the gathered crowd.[65] This cultural expectation of women's accomplishments can be found through Austen's commentary:

> no one can be really esteemed accomplished who does not greatly surpass what is usually met with. A woman must have a thorough knowledge of music, singing, drawing, dancing, and the modern languages, to deserve the word, and besides all this, she must possess a certain something in her air and manner of walking, the tone of her voice, her address and expressions, or the word will be but half-deserved.[66]

Victorian Tea Handbill Advertising Maypole Tea, created by Maypole Dairy Co. Ltd., 1899. Located in William H. Ukers, *All About Tea, Volume II.* (*New York: The Tea and Coffee Trade Journal Company, 1935.*), 298.

Lipton's exotic Ceylon tea, September 17, 1892, created by Thomas J Lipton Co. British Library P.P.7611.

Tea on a Terrace, painting by Frédéric Soulacroix, circa 1900. Wikipedia Commons, Q1931234

Woman of gentility and leisure drinking tea, Lithograph by E.B. and E.C. Kellogg, circa 1854. Smithsonian National Museum of American History, DL.60.2259

Demonstration of one's accomplishments allowed women to set themselves up for praise, and to allow young men to chivalrously escort them to their musical instrument.[67] As an event, tea allowed young women to present themselves as worthy of courtship by eligible men. This became an extension of women's role at the tea table. Defining themselves through their wealth, social status, and accomplishments, women of the upper and middle classes used tea culture of the period to assert themselves in their assigned role.

Presentation served as much purpose as the tea and food did at these gatherings. Queen Victoria regularly held afternoon teas and garden parties, where her daughters served tea and cakes to ensure no guest left hungry or wanting for anything.[68] In *Jane Eyre*, Charlotte Brontë described a tea table in the following way: "How pretty, to my eyes, did the china cups and bright teapot look .... How fragrant was the steam of the beverage .... We feasted that evening as on nectar and ambrosia."[69] In Brontë's work, a tea table expressed luxury and comfort for the senses. Women who presided over beautiful tea tables symbolized the comfort and nourishment of home.[70] Due to its association with comfort, tea also served as a healing beverage. This can be best viewed in *Wuthering Heights*, when one woman declares herself dying of heartbreak; it is only when tea arrives that she revives her spirits and determines not to die over a man who did not love her.[71] Beyond healing, tea provided women with the opportunity to express their personality through decorations at the tea table.[72] These expressions of detail gave a personal touch to each tea party.

In the novel *North and South,* Elizabeth Gaskell also incorporated the idea that tea served the senses, through beautiful details that commended the hostess to her guests:

> Behind the door was another table, decked out for tea, with a white tablecloth, on which flourished the cocoa-nut cakes, and a basket piled with oranges and ruddy American apples, heaped on leaves. It appeared to Mr. Thornton that all these graceful cares were habitual to the family; and especially of a piece with Margaret.[73]

These expressions of detail gave a personal touch to each tea party. Through the personalization of the tea table, tea came to be viewed as an extension of the woman of the house among all classes. Bram Stoker utilized this in *Dracula* in 1880. "Mrs. Harker gave us a cup of tea, and I can honestly say that, for the first time since I have lived in it, this old house seemed like home."[74] In stark contrast to that domestic scene is Jonathan Harker's early visit to Castle Dracula, where no tea or women are present. Harker's discomfort in the environment and the absence of tea, representing the honest comfort of home, can later be linked to Dracula's true nature. Stoker's gothic tale of moral good versus evil employed common symbolism of the 1880s, evoking comforting connections to the morality of a British home. Mary Wollstonecraft associated women's morality with society's concern over reputation; women who displayed moral reputations practiced morality at home.[75] The connection between woman's morality and the home naturally progressed to include tea.

As middle-class women remained within the home to provide comfort for the family, they became a shining example for many among the temperance movement during the early Victorian period, between 1830 and 1850.[76] The link between home and tea existed long before temperance societies; however, this movement altered public perception toward the morality of tea. Tea offered an alternative to alcohol.[77] Connecting the primarily feminine association of tea and home gave the movement ammunition to fight against public houses, where ale and gin served to destroy families.[78] Using tea's feminine association to respectability provided a clear message that drinking tea rather than alcohol brought a person gentility. The gender association between tea and women existed from its arrival to England, in contrast to the alcoholic beverages that were called "men's prerogative."[79] Temperance societies promoted tea as a healthy and moral alternative to gin and ale.[80] As tea's association with morality grew in acceptance, public tea houses opened across Britain; these establishments provided women with the ability to sit in a public location and enjoy refreshments without a chaperone or concern for their reputations.[81] By the late 1880s, tearooms in London opened under the management and ownership of women, catering to the tea needs of women.[82]

Tea became an event around which women organized their social lives. These events gave women opportunity to interact with eligible men and friends. The association between tea and morality allowed women to promote their personal reputations, and therefore morality, in drinking the beverage in society. It also gave the temperance movement an alternative to alcohol, started by and with women, and the opportunity to demonstrate women's activism. Arguably the most influential beverage in British history, tea inspired a nation to colonize Asia to maintain their drinking habit and to adapt their manners and culture to include it. "Since the introduction of tea into England, but more especially since the British public has patronised it, a marked improvement characterises the tone and manners of Society."[83] The transition to Indian tea specifically targeted women, as merchants recognized that women were the primary audience for home purchases. Acquiring Indian tea allowed women to support imperial ideals and publicly align themselves with a political cause before women were allowed to vote. Although enjoyed by both genders across the levels of British society, it was specifically the women of the upper classes that tied a strong portion of their identity to tea culture. The rituals and etiquette surrounding society's tea drinking allowed women greater freedom in socialization and provided them the opportunity to demonstrate their patriotism. Tea uniquely afforded women opportunities to develop their identities across three dimensions of British society.

Bibliography

Austen, Jane. *The Collected Novels of Jane Austen*. New York: Random House, 1964.

Brontë, Charlotte. *Jane Eyre: An Autobiography*. London, UK: Service & Paton, 1897.

———. *Villette*. Oxford: Oxford University Press, 1853.

Brontë, Emily. *Wuthering Heights*. New York: Penguin Books, 2003.

Congreve, William. *The Way of the World: A Comedy*. London, UK: Jacob Tonson, 1706.

Day, Samuel Phillips. *Tea, Its Mystery and History*. London, UK: Simpkin, Marshall & Co., 1878. https://babel.hathitrust.org/cgi/pt?id=uc2.ark:/13960/t6d21vv94&view=1up&seq=66.

Ellis, Sarah Stickney. *The Women of England, Their Social Duties, and Domestic Habits*. New York: D. Appleton & Co., 1839. https://babel.hathitrust.org/cgi/pt?id=nyp.33433075964183&view=1up&seq=13.

Fromer, Julie. *Necessary Luxury: Tea in Victorian England*. Athens, OH: Ohio University Press, 2008. https://ebookcentral.proquest.com/lib/apus/detail.action?docID=1743704.

Gaskell, Elizabeth Cleghorn. *North and South*. London, UK: Walter Scott, 1854.

Gray, Sally Helvenston. "Searching for Mother Hubbard: Function and Fashion in Nineteenth-Century Dress." *Winterthur Portfolio* 48, no. 1 (2014): 29–74. https://www.jstor.org/stable/10.1086/676031.

Guerty, P.M., and Kevin Switaj. "Tea, Porcelain, and Sugar in the British Atlantic World." *Organization of American Historians Magazine of History* 18, no. 3 (2004): 56–59. https://www.jstor.org/stable/25163685.

Hartley, Florence. *The Ladies' Book of Etiquette, and Manual of Politeness: A Complete Handbook for the Use of the Lady in Polite Society*. Boston: Lee and Shepard, 1873. https://archive.org/details/bookofetiqladies00hartrich.

Harvey, Karen. "Barbarity in a Teacup? Punch, Domesticity and Gender in the Eighteenth Century." *Journal of Design History* 21, no. 3 (2008): 205–21. https://www.jstor.org/stable/25228589.

Himes, Amanda E. "Reading the Tea Leaves: Sanditon's Reflection of English Identity." *Persuasions On-line* 38, no. 2 (2018). http://jasna.org/publications/persuasions-online/volume-38-no-2/himes/.

Humphry, C.E. *A Word to Women*. London, UK: James Bowden, 1898.

———. *Manners for Men*. London, UK: James Bowden, 1897. https://archive.org/details/mannersformen01humpgoog/page/n5.

Kowaleski-Wallace, Beth. "Women, China, and Consumer Culture in Eighteenth-Century England." *Eighteenth Century Studies* 29, no. 2 (1995/1996): 153–67. https://www.jstor.org/stable/30053280.

Leslie, Eliza. *The Behavior Book: A Manual for Ladies.* Philadelphia: Willis P. Hazard, 1855.

Pettigrew, Jane, and Bruce Richardson. *A Social History of Tea: Tea's Influence on Commerce, Culture and Community.* Danville: Benjamin Press, 2014.

Rappaport, Erica. *A Thirst for Empire: How Tea Shaped the Modern World.* Princeton: Princeton University Press, 2017.

Sigmond, G.G. *Tea; Its Effects, Medicinal and Moral.* London, UK: Longman, Orme, Brown, Green & Longmans, 1839. https://babel.hathitrust.org/cgi/pt?id=uc2.ark:/13960/t8kd21s0w&view=1up&seq=13.

Stoker, Bram. *Dracula.* New York: Grosset & Dunlap, 1897.

Strickland, Agnes. *Lives of the Queens of England from the Norman Conquest.* Edited by Caroline G. Parker. New York: Harper & Brothers, 1882.

Thackeray, William Makepeace. *The History of Pendennis.* London, UK: Smith, Elder & Co., 1898.

Tombs, Robert. *The English and Their History.* New York: Vintage Books, 2014.

Ukers, William H. *All About Tea, Volume II.* New York: The Tea and Coffee Trade Journal Company, 1935.

———. *The Romance of Tea: An Outline of Tea and Tea-Drinking Through Sixteen Hundred Years.* New York: Alfred A Knopf, 1936.

Vickery, Amanda. *The Gentleman's Daughter: Women's Lives in Georgian England.* New Haven: Yale University Press, 2003.

Victualler's Tea Association. *A History of the Sale and Use of Tea in England.* London, UK: Liscensed Victualler's Tea Association, 1870.

Waller, Edmund. *The Poems of Edmund Waller.* Vol. II. London, UK: A.H. Bullen, 1901.

Wollstonecraft, Mary. *A Vindication of the Rights of Woman.* London, UK: Walter Scott, 1792. https://babel.hathitrust.org/cgi/pt?id=umn.31951p01113190c&view=1up&seq=4.

Woolsey, Sarah Chauncey, and Jane Austen. *The Letters of Jane Austen: Selected from the Compilation of her Great Nephew Edward, Lord Bradbourne.* Boston, MA: Little, Brown, and Company, 1908.

Notes

1 Edmund Waller, "Of Tea, Commended by Her Majesty," in *The Poems of Edmund Waller,* Vol. II (London, UK: A.H. Bullen, 1901), 94.

2 Ibid.

3 Jane Pettigrew and Bruce Richardson, *A Social History of Tea: Tea's Influence on Commerce, Culture, and Community* (Danville: Benjamin Press, 2014), 18.

4 Agnes Strickland, *Lives of the Queens of England from the Norman Conquest,* ed., Caroline G. Parker, (New York: Harper & Brothers, 1882), 479.

5 Pettigrew and Richardson, *A Social History of Tea,* 17.

6 Ibid.

7 Ibid.

8 Ibid.

9 Ibid.

10 William Congreve, *The Way of the World, A Comedy* (London, UK: Jacob Tonson, 1700), 46.

11 Julie Fromer, *Necessary Luxury: Tea in Victorian England* (Athens, OH: Ohio University Press, 2008), 12, https://ebookcentral.proquest.com/lib/apus/detail.action?docID=1743704.

12 Ibid.

13 Pettigrew and Richardson, 35.

14 Robert Tombs, *The English and Their History* (New York: Vintage Books, 2014), 339.

15 P.M. Guerty and Kevin Switaj, "Tea, Porcelain, and Sugar in the British Atlantic World," *Organization of American Historians Magazine of History* 18, no. 3: 58, https://www.jstor.org/stable/25163685.

16 Victualler's Tea Association, *A History of the Sale and Use of Tea in England* (London, UK: Licensed Victualler's Tea Association, 1870), 8.

17 Pettigrew and Richardson, 17.

18 Ibid.

19 Jane Austen, *Pride and Prejudice* (New York: Random House, 1964), 253.

20 Sarah Stickney Ellis, *The Women of England, Their Social Duties, and Domestic Habits* (New York: D. Appleton & Co., 1839), 265, https://babel.hathitrust.org/cgi/pt?id=nyp.33433075964183&view=1up&seq=13.

21 Amanda Vickery, *The Gentleman's Daughter: Women's Lives in Georgian England* (New Haven: Yale University Press, 2003), 30.

22 Ibid., 23-24.

23 Ibid., 147.

24 Austen, *Pride and Prejudice*, 240.

25 Ibid.

26 Jane Austen, *Emma* (New York: Random House, 1964), 774.

27 William Makepeace Thackeray, *The History of Pendennis* (Project Gutenberg,2009), 898.

28 Vickery, *The Gentleman's Daughter*, 568.

29 Ibid.

30 Victualler's Tea Association, *A History of the Sale and Use of Tea in England*, 5.

31 Pettigrew and Richardson, 18.

32 Waller, "Of Tea, Commended by Her Majesty," 94.

33 G.G. Sigmond, *Tea; Its Effects, Medicinal and Moral* (London, UK: Longman, Orme, Brown, Green & Longmans, 1839), 143, https://babel.hathitrust.org/cgi/pt?id=uc2.ark:/13960/t8kd21s0w&view=1up&seq=13.

34 Jane Austen, *The Letters of Jane Austen: Selected from the Compilation of her Great Nephew Edward, Lord Bradbourne,* comp. Sarah Chauncey Woolsey (Boston, MA: Little, Brown, and Company, 1908), 250.

35 Sigmond, *Tea,* 1.

36 Erica Rappaport, *A Thirst for Empire: How Tea Shaped the Modern World* (Princeton: Princeton University Press, 2017), 126.

37 Ibid.

38 Ibid.,120.

39 Sigmond, 3.

40 Rappaport, *A Thirst for Empire*, 85.

41 Ibid.

42 Ibid., 87.

43 Ibid.

44 Ibid., 166.

45 Ibid.

46 Ibid., 137.

47 Ibid.

48 William H. Ukers, *The Romance of Tea: An Outline of Tea and Tea-Drinking Through Sixteen Hundred Years* (New York: Alfred A Knopf, 1936), 180.

49 Rappaport, 176.

50 Charlotte Brontë, *Villette* (Project Gutenberg, 2003), 412–13.

51 Ukers, *The Romance of Tea*, 169.

52 Fromer, *Necessary Luxury*, 35.

53 Mrs. C.E. Humphry, *Manners for Men*, (London, UK: James Bowden, 1897), 80.

54 Rappaport, 58.

55 Ibid.

56 Pettigrew and Richardson, 56.

57 Austen, *The Letters of Jane Austen.* 30.

58 Jane Austen, *Mansfield Park* (New York: Random House, 1964), 601.

59 Ibid.

60 Ibid., 703.

61 Sally Helvenston Gray, "Searching for Mother Hubbard: Function and Fashion in Nineteenth-Century Dress," *Winterthur Portfolio* 48, no. 1 (2014): 48, https://www.jstor.org/stable/10.1086/676031.

62 Eliza Leslie, *The Behavior Book: A Manual for Ladies* (Philadelphia: Willis P. Hazard, 1855), 80–82.

63 Gray, "Searching for Mother Hubbard," 47.

64 Humphry, *Manners for Men*, 80.

65 Ibid.

66 Austen, *Pride and Prejudice*, 253.

67 Humphry, *Manners for Men*, 80.

68 Pettigrew and Richardson, 140.

69 Charlotte Brontë, *Jane Eyre: An Autobiography* (Project Gutenberg, 2007), 178–80.

70 Fromer, 66.

71 Emily Brontë, *Wuthering Heights* (Project Gutenberg, 2007), 248.

72 Elizabeth Gaskell, *North and South* (Project Gutenberg, 2011), 186.

73 Ibid.

74 Bram Stoker, *Dracula*, (Project Gutenberg, 2013), 521.

75 Mary Wollstonecraft, *A Vindication of the Rights of Woman* (London, UK: Walter Scott, 1792), 196, https://babel.hathitrust.org/cgi/pt?id=umn.31951p01113190c&view=1up&seq=4.

76 Rappaport, 62.

77 Ibid., 61

78 Ibid.

79 Congreve, *The Way of the World*, 47.

80 Rappaport, 61.

81 Ibid., 204.

82 Ibid., 167.

83 Samuel Phillips Day, *Tea, Its Mystery and History* (London, UK: Simpkin, Marshall & Co., 1878), 60, https://babel.hathitrust.org/cgi/pt?id=uc2.ark:/13960/t6d21vv94&view=1up&seq=66.

Unintentional History Makers: Evolution of Feminist Historiography

Tara Dyson
American Public University

Abstract

What is now considered the early feminist movement initially began in simplicity as women's endeavour to gain social, political, and educational equality. Slowly emerging into Western civilization during the eighteenth and nineteenth centuries, the movement birthed new ideas regarding American society, which ultimately perpetuated continual change throughout history. Women such as Abigail Adams, Mary Wollstonecraft, Lucy Aikin, Maria Edgeworth, Harriet Martineau, Elizabeth Cady Stanton, Susan B. Anthony, and Virginia Woolf produced literature that transversed the gender limitations of their times, endeavoring to achieve equality in varying aspects of American culture. Their documentation requires acknowledgement of women's roles throughout history, their contributions to it, and their evolution of change within historiography itself.

Keywords: feminism, historiography, American Revolution, gender, vindication, socialism, sociology, Irish Rebellion

Creadores de historia no intencionales: evolución de la historiografía feminista

Resumen

Lo que ahora se considera el primer movimiento feminista comenzó inicialmente con simplicidad como el esfuerzo de las mujeres por lograr la igualdad social, política y educativa. Lentamente emergiendo en la civilización occidental durante los siglos XVIII y XIX, el movimiento dio a luz nuevas ideas sobre la sociedad estadounidense que finalmente perpetuaron el cambio continuo a lo largo de la historia. Mujeres como Abigail Adams, Mary Wollstonecraft, Lucy Aikin, Maria Edgeworth, Harriet Martineau, Elizabeth Cady Stanton, Susan B. Anthony y Virginia Woolf produjeron literatura que atravesaba las limitaciones de género de su tiempo tratando de lograr la igualdad en diversos aspectos de la cultura estadounidense. . Su documentación requiere el reconocimiento de los roles de las mujeres a lo largo de la historia, sus contribuciones y su evolución del cambio dentro de la historiografía.

Palabras clave: feminismo, historiografía, revolución americana, género, reivindicación, socialismo, sociología, rebelión irlandesa

无意的历史缔造者：女性主义历史学演变

摘要

当下被视为早期的女性主义运动最初是从女性争取获得社会平等、政治平等和教育平等开始的。该运动在18世纪和19世纪缓慢进入西方文明，诞生了有关美国社会的新观点，最终让持续的变革在历史中进行。例如阿比盖尔·亚当斯、玛丽·沃斯通克拉夫特、露西·艾金、玛丽亚·埃奇沃思、哈里特·马蒂诺、伊丽莎白·卡迪·斯坦顿、苏珊·安东尼、以及弗吉尼亚·伍尔夫等女性创作的一系列文学突破了当时的性别限制，努力实现美国文化不同方面的平等。有关她们的文件记载要求承认女性在历史中发挥的作用，女性对历史的贡献，以及女性在历史学中的变化发展。

关键词：女性主义，历史学，美国革命，性别，辩护，社会主义，社会学，爱尔兰叛乱

Unintentional History Makers: Evolution of Feminist Historiography

The role of women in history has been undervalued, if not entirely overlooked, in past centuries due to the cultural restrictions of their role within society. However, the emergence of the feminist movement in Europe in the nineteenth century had a significant global impact on women in the twentieth and twenty-first centuries. As the feminist movement slowly emerged within Western civilization during the nineteenth century, it birthed new ideas about the role of women in society, education, and politics, impacting the study of historiography. Women have now been studied as historians, and in some cases, as pioneers of various historiographical schools of thought. In centuries prior, wherein women's personal virtue was held in higher esteem than their political or academic achievement or individual thought, there were those who dared to write history from their perspectives, heedless of society's conventions. Women who dared to do so inadvertently gave voice and credence to the evolution of feminist historiography.

The development of women as historians has no definite beginning. To pinpoint one individual whose sole influence significantly impacted those who have led the charge of feminism in itself and in relation to history is simply impossible, as there are many who are worthy of such esteem. Many modern historians of the later centuries attribute the musings of Virginia Woolf as an impetus for consideration of women as historians. In her *A Room of One's Own* (1929), she ponders the life of women in the past by questioning,

> What one wants, I thought- and why does not some brilliant student at Newnham or Girton supply it?—is a mass of information; at what age did she marry; how many children had she as a rule? ... All these facts lie somewhere, presumably, in parish registers and account books; the life of the average Elizabethan woman must be scattered about somewhere, could one collect it and make a book of it. It would be ambitious beyond my daring, I thought, looking about the shelves for books that were not there, to suggest to the students of those famous colleges that they should rewrite history, though I own that it often seems a little queer as it is, unreal, lop-sided; by why should they not add a supplement to history?[1]

In her musing regarding the society of Elizabethan England and in her study of Shakespeare, she recognized a lack of women in its telling. Although she was not a historian, Woolf's influence upon those who would later reference her works as those of historical significance altered the overall perception of her as simply a literaturist, therefore thrusting her into the unintended role of historian. She spent much time pondering history; specifically regarding women's role in it. In *To the Lighthouse* (1927), she notes that "the past is beautiful because one never realises an emotion at the time. It expands later, and thus we don't have complete emotions about the present, only about the past."[2] Her eloquent expression of history inspired women of future generations. Ironically, she describes later that the role of women in history is "superfluous and indispensable."[3] Although Woolf's musings of women in history extended only to her writings, as she never intentionally took up the study of history, she challenged others to do so. Those whose lives were affected by Woolf's provocation, who took upon such an endeavor as she challenged, would later prove to be worthwhile in their mark upon the study of history. She redefines a traditional masculine history to record those details of suppressed individuals, such as women, and requires from the record a more complex, deeper truth. In an endeavor to feminize historiography, women and their writings would come to do just as Woolf proposed.

Until the later twentieth century, history has been predominantly recorded by men, or at least interpreted as having been so. A thorough study into history, however, produces works by women that have recorded life in far more detail than their male counterparts, just as Woolf suggested. Of the American women whose writings have been esteemed for their historical significance is Abigail Adams, wife of the second president, John Adams. Although Adams's intent was not as a historian, she has now in retrospect become hailed as such. In a 1975 study, *Weathering the Storm: Women of the American Revolution,* Elizabeth Evans declares that "The most famous advocate for women's rights was Abigail Smith Adams, wife of John Adams .... Refusing to be an obscure mouthpiece for her husband's views, she influenced many of his political decisions."[4] Adams is most remembered for her suggestion to her husband regarding the composition of the Constitution. In one of her many letters to him she endeavors to persuade him to "remember the ladies, and be more generous and favorable to them than your ancestors. Do not put such unlimited power into the hands of the Husbands. Remember all Men would be tyrants if they could. If particular care and attention is not paid to the Ladies we are determined to foment a Rebellion, and will not hold ourselves bound by any Laws in which we have no voice, or Representation."[5] Prior to the birth of the American nation, she realized the importance of the quality of sexes and the repercussions of ignoring them, which has become a focus of study for many historians in the twenty-first century, most notably Dr. Edith Gelles, whose recent endeavor is editing the Library of America's Letters of Abigail Adams.[6] Of Adams, Gelles states: "[Adams] writes the history of her age from a woman's perspective .... She also displays a rare honesty and accuracy, and she describes feelings as well as events."[7]

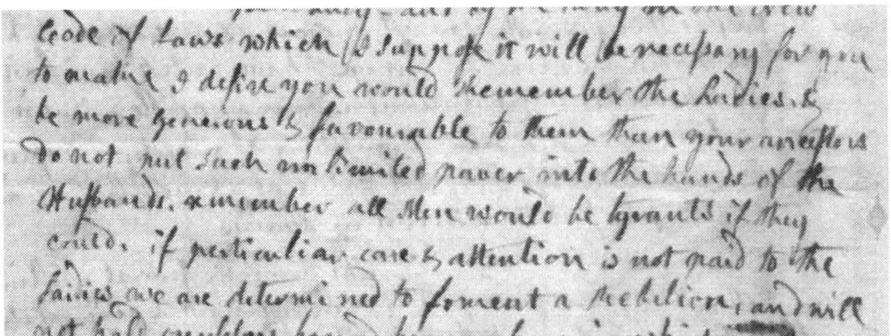
Code of Laws which I suppose it will be necessary for you
to make I desire you would Remember the Ladies, &
be more generous & favourable to them than your ancestors
Do not put such unlimited power into the hands of the
Husbands. Remember all Men would be tyrants if they
could. if perticular care & attention is not paid to the
Laidies we are determined to foment a Rebelion, and will
not hold ourselves bound by any Laws in which we have
no voice, or Representation

Letter from Abigail Adams to John Adams, March 31 to April5, 1776 [electronic edition]. *Adams Family Papers: An Electronic Archive.* Massachusetts Historical Society.

Adams lived during a time in which women did not have legal rights and under the laws of the time, women were entities belonging to their husbands, and any property owned by them was essentially the property of their husbands.[8] Much of these beliefs were borne out of the Christian values that had so affected prior generations in the formation of their beliefs about women's roles. Of the many verses in the Holy Bible regarding women's roles, that which was exacted in the most severe of interpretations is I Timothy 2:11–12, which states: "Let the woman learn in silence with all subjection. But I suffer not a woman to teach, nor to usurp authority over the man, but to be in silence." Its projection was taken not in the original text, but leant itself to the subjection of women under the laws of the time. Abigail Adams defied such restraints both legally and ethically throughout much of her lifetime. Through her correspondence, public records, and oral histories, historians have discovered that during her lifetime, Adams continually circumvented the laws by purchasing land and assets in the names of her husband and sons, only to use the revenue from such to build her own personal wealth.[9] After having accumulated much wealth during her lifetime due to successful investment, on her deathbed, she composed a handwritten will in which she bequeathed her money and possessions to her female relations. Adams left little to her two sons, and the remainder was left to daughters, nieces, granddaughters, and even one female servant.[10] In leaving her possessions to women, she subtly

Abigail Smith Adams, 1879; photograph by Chester A. Lawrence of painting by Gilbert Stuart. Library of Congress Prints and Photographs Division.

proclaimed her beliefs of their rights as citizens of America. Adams's forthright manner might have been squashed by some men of the time. John Adams had it within his legal right to overturn his wife's desires. However, in knowing her wishes to empower those to whom she had given her possessions, he legalized her will and, in carrying it out, therefore acknowledged her rights.[11] Abigail Adams's life and actions is a continued testament to the lives of those men whose names have resonated in history as the "Founding Fathers" of America. Although she is not listed among them, her influence is apparent to those who study her correspondence. Much like Thomas Jefferson, whose correspondence has long been considered a cornerstone for historical study, Abigail's historiography reconceptualizes the history of the eighteenth century in America.

Even before these writings were considered foundational to the study of history, as they are so credited today, the European Enlightenment produced prominent women writers who struggled against the preconceived ideas of their sex. Of those whose ideas have inspired modern women historians, there is Mary Wollstonecraft (1759–1791). Wollstonecraft cemented her feminist and socialism ideals in the late eighteenth century. As a testament to her values, she adopted the "unladylike" profession of writing beyond that of as a mere hobby. In doing so, she embodied those principles of feminism she held so dear and expressed at every opportunity.[12] Her eloquence was found in her ability to write to the heart of the matter without the use of fancy wording so as to exact a concise conscription of the events at hand. Her *A Vindication of the Rights of Man* (a response to Edmund Burke's *Reflections on the Revolution in France*) is a historical document primarily of ideals often associated today with socialism.[13] Throughout the work, she references John Locke's ideas of working toward the ownership of one's property so that the poor might become wealthier and that the aristocracy be diminished in the effort.[14] Wollstonecraft's documentation of the French Revolution provides insight into the daily life of the common people in the lower classes in a voice much unlike the male historians of her time.

Through Wollstonecraft's writings, modern Marxist feminist historians have redefined the social expectations of women. These expectations go beyond the mode of production usually studied to envelope the modes of reproduction.[15] Feminist historians such as Dr. Jane More argue that Wollstonecraft broke barriers long before her time in these areas. According to More, Wollstonecraft "aims to show that women are less different from men than popular moralists and educationalists would have it."[16] Furthermore, Wollstonecraft exposed the nature of politics in Europe during the Enlightenment while also shedding light on the life of a single woman during this time by divulging her own experiences.[17] Historian Brian Vickers argues that "modern social history encourages us to see the relation between the sexes in early modern period in terms of 'more/less' rather than 'all/none.'"[18] Unfortunately, the relation of such was overshadowed by society's view of women at the time. Wollstonecraft's being a woman led reviewers of her works to attribute her beliefs simply to the "passion" of women, and they were therefore considered irrational translations of the revolution. Although she has since been hailed as a great pioneer of socialist historiography, her personal life altered the perception her contemporaries had of her in such a way that she never gained the reputation she has been afforded in modern times.[19] Her aforementioned works, as well as *A Vindication of the Rights of Woman: With Strictures on Political and Moral Subjects* (1792), wherein she further philosophizes about the subject of

Mary Wollstonecraft; engraving by James Heath and John Opie. Library of Congress, Lot 13301, no. 108.

women in both the political and social arenas, have become foundational in studying the history of the eighteenth century.

Following in the footsteps of Wollstonecraft was the Unitarian feminist writer, Lucy Aikin (1781–1864). Aikin possessed "a profound conviction in the capacity of right reason to discover the truths both of human nature and the physical world, the belief that souls have no sex and that moral virtues and intellectual capacities are available to both sexes in equal measure, and the idea that history is a process that moves by uneven developments towards higher levels of human civilization."[20] Her widely regarded *Epistles on Women* (1810) was composed in response to Mary Wollstonecraft's plea to women to disengender history. Aikin argued that history is in fact "impartial to gender," and surveyed history from the time of the savages to those people civilized in the western hemisphere.[21] As the title suggests, she had an intended public audience. The work itself is a 1,200 line poem in which Aikin investigates numerous historical events throughout centuries of world history, particularly seventeenth century England. In preparation for the composition, Aikin studied the works of Montesquieu; his theory of the separation of powers is evident throughout her epilogue.[22] She personifies various values in gender, those values of a positive nature are embodied as female, and those negatively perceived are male.[23] Aikin's perspective of the history has been reflected through modern feminist historians. Historian Judith Bennett expounds

Lucy Aikin; pencil portrait by Edmund Aikin (1780–1820), Hampstead Parish Church Gallery.

upon Aikin's opinion, showing that through the social standards of imbalance of wages, suppressed public voice, political effrontery, and prevention of education, women suffered stagnation. While men enjoyed a time of great enlightenment and success, women continued under the bonds of their station as subservient helpers to their counterparts.[24] Throughout Aikin's poem, she evidently leans toward a feminist bias. Its utilization as a historical document and insight into history through the female perspective is notable for historical study. Lucy Aikin's work is highly regarded by feminist historians, as her perspective provides groundwork for understanding world history and a woman's place within it.[25]

During the time when Adams's, Wollstonecraft's, and Aikin's writings projected a world wherein equality was possible, Irish writer Maria Edgeworth (1768–1849) published her novel *The Absentee* (1812). Through fictional characters, Edgeworth studies the issues of absenteeism through a time period in which the Irish were, once again, at odds with the British. In her novel, she portrays life through realism by "transforming the individual historiographical perspective into a broader, collective historical consciousness."[26] While presented as a fictionalized tale, Edgeworth accurately records the historical issues surrounding absenteeism through the context of the virtues of the people and the justice of the system that had been implemented long before her time. In the study of the context of the tensions between Ireland and Great Britain, one can better understand their causality and effects upon the citizenry of the time period.[27] In many of Edgeworth's writings, she sought out answers to society's issues through questioning religion and gender, and their influence upon one another. Many of her novels were based solely upon her observations, but dictated with a concise method.[28] Edgeworth provides an accurate historical depiction and explanation of absenteeism from both a personal stance and that of the members of the society involved.[29] According to historian Deborah Weiss:

> Maria Edgeworth, on the other hand, has never been considered a radical thinker despite the fact that she was as critical as Wollstonecraft of her society's gender codes and their effect on the moral and intellectual lives of women. At the level of theory, as well, Edgeworth resembled Wollstonecraft in her rejection of the period's essentialist understanding of gender. Like Wollstonecraft, Edgeworth took Enlightenment concepts of the cultural formation of the individual—ideas used by radical male thinkers such as William Godwin to argue for the universal equality of "mankind"—and applied these concepts to the formation of feminine identity.[30]

Edgeworth's contribution through the application of her beliefs inarguably allowed historians to follow her in contextualizing the roles of women during the political turmoil of the nineteenth century.

Maria Edgeworth, 1841; daguerreotype by Richard Beard. National Portrait Gallery, P5.

During the time in which Edgeworth made her mark in history, the British social theorist Harriet Martineau (1802–1876) debuted her observations along the same school of thought. She has been conferred the title of "first woman sociologist" by modern sociological and feminist historians.[31] Martineau's sociological study *Society in America* (1837) was written after a two-year tour of America in 1834. In it, she completes her observations about the realities of American culture during a time in which states' rights, Indian removal, the Texas Revolution, and the nullification crisis were on the forefront of American society's political issues. She notes areas that had been previously neglected by other authors.[32] Martineau asserts, "for the character of people has such a complexity of aspect, that even the honest observer knows not always, not perhaps after long inspection of it, what to determine regarding it."[33] She admits that her observations of American society are limited and can only be perceived in regard to her own understanding of it, although her objectivity is easily noted. During her tour, she documented the political history of America while also noting the social and economic situation.[34] Martineau goes on to say, "The United States have indeed been useful in proving these two things, before held impossible; the find of a true theory of government, by reasoning of the principles of human nature as well as from the experience of government; and the capacity of mankind for self-government."[35] Even in her 1837 publication, she conceptualizes and expounds upon the "things" of human behavior as that cannot otherwise be defined. These "things" were later attributed to Émile Durkheim in his *The Rules of Sociological Method* (1895), wherein he suggests: "The first and fundamental rule [of sociology] is to consider social facts as things ... a social fact is every way of acting which is capable of exercising an external constraint upon the individual."[36] Prior to the published works of her male counterparts, including Karl Marx, Émile Durkheim, or Max Weber, Martineau observed and examined the social division of classes in American culture through religion, character, public institutions, and social status, particularly that of women.[37] The methodology she used in her analysis of classes was groundbreaking in that she depended upon logic and scientific observation rather

Harriet Martineau, 1861; albumen print by Camille Silvy. National Portrait Gallery, 33.

than sheer emotion, as many women were wont to do.[38] Her implementation of such later influenced sociologist historians in approaching history using a logical means, which would continue to evolve into the Enlightenment period.[39] Through Martineau's efforts and the utilization of her objectivity, social historians now study history from the human experience while eradicating the simple study of events and dates.[40] Her assertions set precedent for modern day historians, both men and women, in the area of inclusion of women and social class within historical study.

During the time that Martineau traveled and observed American society, feminist leaders were emerging in the women's suffrage movement: women such as Elizabeth Cady Stanton (1815–1902) and Susan B. Anthony (1820–1906). It is due to their strenuous efforts and sacrifices that the women's movement is now historically studied in correlation to the economy, political movements, and the reformation of education.[41] In a time when women lived in a society that prevented them property rights, professional careers, public speaking, economic investing, and many more rights, these women emerged as spokespersons for the "certain unalienable rights" that their Founding Fathers had set forth a century prior.[42] During the Seneca Falls Convention of 1848, Stanton and other members used the Declaration of Independence as the basic template upon which they drafted the Declaration of Sentiments, which declared that "women under this government ... demand the equal station to which they are entitled."[43] Her collaboration with Susan B. Anthony formed a framework upon which feminists and women's suffragists continued to build a compelling argument regarding the rights of women in early twentieth century America. In a letter to Anthony, fellow suffragist Lydia Becker succinctly summarizes their efforts: "Beyond the material gains in legislation, we find a general improvement in the tone of feeling and thought toward women—an approach, indeed, to the sentiment recently expressed by Victor Hugo, that as man was the problem of the eighteenth century, woman is the problem of the nineteenth century. May our efforts to solve this problem lead to a happy issue."[44]

The Emancipation of Women; painting by Hildreth Meière,photographed by Peter A. Juley & Son, Smithsonian American Art Museum Archives and Special Collections, Jul J0012534.

Calling upon the writings of the Founding Fathers and echoing the plea of Abigail Adams many years before, these women sought to change history through their involvement in politics. While the success of their endeavor was not obtained during either woman's lifetime, their legacy continued into the turn of the century. Historians now investigate the correspondence of these women as important pieces of evidence for the suffragist movement, which is considered a foundation for what would later become the feminist movement.

While these women's writings left a deep impression upon the feminist historiographical school of thought, the term "feminist" would not have been introduced into their vocabulary as it was not used until the 1890s in France (derived from the French word *femme*); even then it did not catch on in popular usage until far later in the twentieth century.[45] In the latter half of the twentieth century, the word "feminist" found its application to the study of history as the American feminist movement gained more momentum in education and politics.[46] During this time, a conscious effort was made by the academic world to include women's roles in history. History became a tool of feminism for celebrating the past accomplishments of these forerunners and a continuing plea to their posterity to pursue it as a greater goal.[47] The conscientious efforts of feminists who once transversed through societies in which the ideals of equality for women were unpopular have now elevated women of the modern world to a position of equality in areas that were never once thought possible. Through the unintentional efforts of women who documented history to those who painstakingly fought for it, history now acknowledges women's role in it, their literary contributions to it, and the importance of understanding it.

Bibliography

Adams, John, Abigail Adams, and Frank Shuffelton. *The Letters of John and Abigail Adams.* New York: Penguin Books, 2004.

Aikin, Edmund. *Lucy Aikin.* Hampstead Parish Church Private Collection. Simon Fraser University. Vancouver, Canda. http://www.sfu.ca/~mnl/aikin/Site/Biography.html.

Beard, Richard. *Maria Edgeworth.* 1841. National Portrait Gallery, P5. https://www.npg.org.uk/collections/search/portrait/mw02020/Maria-Edgeworth.

Bennett, Judith M.. *History Matters: Patriarchy and the Challenge of Feminism.* Philadelphia: University of Pennsylvania Press, 2006.

Burke, Peter. *New Perspectives on Historical Writing.* University Park: Pennsylvania State University Press, 1992.

Durkheim, Émile. *The Rules of Sociological Method.* 8th ed. Collier-Macmillan: Free Press, 1964.

Evans, Elizabeth. *Weathering the Storm: Women of the American Revolution.* New York: Scribner, 1975.

Heath, James, and John Opie, engraver. *Mary Wollstonecraft,* half-length portrait, facing left. New York: Rockwood photographer. Library of Congress Prints and Photographs Division, Lot 13301, no. 108. https://www.loc.gov/item/2004680088/.

Hill, Michael R., and Susan Drysdale. *Harriet Martineau: Theoretical and Methodological Perspectives.* New York: Routledge, 2001.

Holton, Woody. "Abigail Adams' Last Act of Defiance." *American History* 45, no. 1 (April 2010): 56-61.

"Letter from Abigail Adams to John Adams, 31 March - 5 April 1776" [electronic edition]. *Adams Family Papers: An Electronic Archive.* Massachusetts Historical Society. http://www.masshist.org/digitaladams/.

Martineau, Harriet. *Society in America.* New York: Saunders and Otley, 1837.

Meière, Hildreth. *The Emancipation of Women*. 1937. Photographed by Peter A. Juley & Son. Archives and Special Collections, Jul J0012534, Smithsonian American Art Museum. https://siris-juleyphoto.si.edu/ipac20/ipac.jsp?&profile=julall&source=~!sijuleyphotos&uri=full=3100001~!12534~!0#focus.

Mellor, Anne K. "Telling Her Story: Lucy Aikin's Epistles on Women (1810) and the Rewriting of Western History." *Women's Studies* 39, no. 5 (July 2010): 430-450.

More, Jane. *Mary Wollstonecraft*. Oxford: Oxford University Press, 2018.

Moses, Claire Goldberg. ""What's in a Name?" On Writing the History of Feminism." *Feminist Studies* 38, no. 3 (2012): 757-79. www.jstor.org/stable/23720210.

Nash, Julie. *New Essays on Maria Edgeworth*. Aldershot, UK: Ashgate Publishers, 2006.

"News Release." *Abigail Adams: History from a Woman's Perspective*. Stanford University, (December 1992). news.stanford.edu/pr/92/921201Arc2028.html.

Pedersen, Joyce Senders. "Mary Wollstonecraft: A Life in Past and Present Times." *Women's History Review* 20, no. 3 (July 2011): 423-436.

Pennell, Elizabeth Robins. *Mary Wollstonecraft Godwin*. London, UK: W.H. Allen, 1885.

Scott, Joan Wallach. *Gender and the Politics of History*. New York: Columbia University Press, 1988.

Silvy, Camille. *Harriet Martineau*. 1861. National Portrait Gallery, NPG 33. https://www.npg.org.uk/collections/search/portrait/mw04253/Harriet-Martineau?search=sp&sTet=martineau&rNo=3.

Stanton, Elizabeth Cady. Susan B. Anthony, and Matilda Jocelyn Gage. *History of Woman Suffrage* Vol. 3. New York: Susan B. Anthony, 1886. Ebook. http://www.gutenberg.org/files/28556/28556-h/28556-h.htm.

Stuart, Gilbert. *Abigail Smith Adams*. 1879. Photography by Chester A. Lawrence. Library of Congress Prints and Photographs Division. Control 2004682096. https://lccn.loc.gov/2004682096.

Vickers, Brian. "Francis Bacon, Feminist Historiography, and the Dominion of Nature." *Journal of the History of Ideas* 69, no. 1 (2008): 117-141. https://www.jstor.org/stable/30139670.

Ward, Geoffrey C., Martha Saxton, Ann D. Gordon, and Ellen Carol DuBois. *Not for Ourselves Alone: The Story of Elizabeth Cady Stanton and Susan B. Anthony: An Illustrated History*. New York: A.A. Knopf, 1999.

Weiss, Deborah. "The Extraordinary Ordinary Belinda: Maria Edgeworth's Female Philosopher." *Eighteenth Century Fiction* 19, no. 4 (Summer 2007): 441-461.

Wollstonecraft, Mary. *A Vindication of the Rights of Men*. New York: Cosimo, 2008. Ebook.

Woolf, Virginia. *A Room of One's Own*. San Diego: Harcourt Brace Jovanovich, 1989.

Woolf, Virginia. *To the Lighthouse*. San Diego: Gardners Books, 1987.

Notes

1 Virginia Woolf, *A Room of One's Own* (San Diego: Harcourt Brace Javanovich, 1989), 82.

2 Virginia Woolf, *To the Lighthouse* (San Diego: Gardners Books, 1987), 5.

3 Peter Burke, *New Perspectives on Historical Writing* (University Park: Pennsylvania State University Press, 1992), 50.

4 Elizabeth Evans, *Weathering the Storm: Women of the American Revolution* (New York: Scribner, 1975), 5.

5 John Adams, Abigail Adams, and Frank Shuffelton, *The Letters of John and Abigail Adams* (New York: Penguin Books, 2004).

6 "News Release," *Abigail Adams: History from a Woman's Perspective* (Standord: Stanford University, 1992), l.

7 Ibid.

8 Woody Holton, "Abigail Adams' Last Act of Defiance," *American History* 45, no. 1 (April 2010): 52.

9 Ibid., 57.

10 Ibid., 60.

11 Ibid., 61.

12 Elizabeth Robins Pennell, *Mary Wollstonecraft Godwin* (London, UK: W.H. Allen, 1885), 50.

13 Mary Wollstonecraft, *A Vindication of the Rights of Men* (New York: Cosmo, 2008). Ebook.

14 Pennell, *Mary Wollstonecraft Godwin*, 12.

15 Joan Wallach Scott, *Gender and the Politics of History* (New York: Columbia University Press, 1988), 16.

16 Jane More, *Mary Wollstonecraft* (Oxford: Oxford University Press, 2018), 39.

17 Pennell, 51-53.

18 Brian Vickers, "Francis Bacon, Feminist Historiography, and the Dominion of Nature," *Journal of the History of Ideas* Vol. 69, No. 1 (January, 2008), 120.

19 Joyce Senders Pedersen, "Mary Wollstonecraft: A Life in Past and Present Times," *Women's History Review* 20, no. 3 (July 2011), 426.

20 Anne K Mellor, "Telling Her Story: Lucy Aikin's Epistles on Women (1810) and the Rewriting of Western History," *Women's Studies* 39, no. 5 (July 2010), 430.

21 Ibid., 432.

22 Ibid., 433.

23 Ibid., 440.

24 Judith M.Bennett, *History Matters: Patriarchy and the Challenge of Feminism* (Philadelphia: University of Pennsylvania Press, 2006), 106.

25 Burke, *New Perspectives,* 54.

26 Julie Nash, *New Essays on Maria Edgeworth* (Aldershot, UK: Ashgate, 2006), 175.

27 Ibid., 176.

28 Ibid., 178.

29 Ibid., 179.

30 Deborah Weiss, "The Extraordinary Ordinary Belinda: Maria Edgeworth's Female Philosopher," *Eighteenth Century Fiction* 19, no. 4 (Summer 2007), 441.

31 Michael R. Hill and Susan Drysdale, *Harriet Martineau:Theoretical and Methodological Perspectives* (New York: Routledge, 2001), 6.

32 Ibid., xv.

33 Harriet Martineau, *Society in America* (New York: Saunders and Otley, 1837), ii.

34 Hill and Drysdale, *Harriet Martineau,* 8.

35 Martineau, *Society in America,* 2-3.

36 Emile` Durkheim, *The Rules of Sociological Method.* 8th ed. (Chicago: Collier-Macmillan: Free Press, 1964), 147.

37 Hill and Drysdale, *Harriet Martineau,* 9.

38 Ibid., 12.

39 Ibid., 14.

40 Ibid., 26.

41 Scott, *Gender,* 15.

42 Geoffrey C.Ward et al., *Not for Ourselves Alone: The Story of Elizabeth Cady Stanton and Susan B. Anthony: an Illustrated History* (New York: A.A. Knopf, 1999), viii.

43 Ibid., 39.

44 Elizabeth Cady Stanton, Susan B. Anthony, and Matilda Jocelyn Gage, *History of Woman Suffrage*, Vol. 3 (New York: Susan B. Anthony, 1886), 62. Ebook.

45 Claire Goldberg Moses, "'What's in a Name?' On Writing the History of Feminism," *Feminist Studies* 38, no. 3 (2012): 762.

46 Burke, *New Perspectives,* 43.

47 Bennett, *History Matters,* 6.

Not Fit to Breed: Eugenics in Sweden, 1900 to Present

Susan Danielsson
American Military University

Abstract

In the 1930s, the Swedish government enacted eugenic policies that permitted the forced sterilization of individuals the government deemed unfit to reproduce, often targeting them with accusations of mental illness. When officials passed the Sterilization Acts, they kept socioeconomic benefits in mind, but the eugenics movement in Sweden had deep roots in race-based science. Charles Darwin and his famous works on evolution inspired Swedish scholars to promote social hygiene within their own population, and they used political parties and the elite to push their agenda into social policy. Officials implemented sterilization laws that were intended to improve the gene pool of the Swedish population as a way to ensure the affordability of their welfare system. In the 1950s, the Swedish government started to prioritize the rights and wants of the individual, instead of making them second to the wellbeing of society.

Keywords: Sweden, Scandinavia, eugenics, race, social policy, biology, Nordic, racial hygiene, mental illness

No apto para la raza: eugenesia en Suecia, 1900 hasta el presente

Resumen

En la década de 1930, el gobierno sueco promulgó políticas eugenésicas que permitían la esterilización forzada de aquellos individuos que el gobierno consideraba incapaces de reproducir, a menudo atacándolos con enfermedades mentales. Cuando los funcionarios aprobaron las Leyes de esterilización, tuvieron en cuenta los beneficios socioeconómicos, pero el movimiento eugenésico en Suecia tenía profundas raíces en la ciencia basada en la raza. Charles Darwin y sus famosos trabajos sobre la evolución inspiraron a los académicos suecos en la promoción de la higiene social dentro de su propia población, y utilizaron los partidos políticos y la élite para impulsar su agenda en la política social. Los funcionarios implementaron las leyes de esterilización para mejorar el acervo genético de la población sueca con el fin de garantizar la asequibilidad de su sistema de bienestar. En la década de 1950, el gobierno sueco comenzó a priorizar los derechos y las necesidades del individuo, en lugar de hacer que fueran secundarios para el bienestar de la sociedad.

Palabras clave: Suecia, Escandinavia, eugenesia, raza, política social, biología, nórdico, higiene racial, enfermedad mental

不适合生育：1990年到现在的瑞典优生学

摘要

20世纪30年代，瑞典政府通过了优生政策，允许对那些被政府认为不适合生育的个人进行强制绝育，这些个人通常患有精神疾病。当政府官员通过绝育法时，他们考虑的是社会-经济利益，但瑞典的优生运动根植于基于种族的科学。查尔斯·达尔文及其在进化论方面的名著启发了瑞典学者在本国人口中推动社会卫生学，并且他们利用政党与精英推动其议程融入社会政策。政府官员落实绝育法以提升瑞典人口基因库，进而确保其福利体系的可负担性。瑞典政府在20世纪50年代开始优先考虑个人的权利和需求，而不是让其次之于社会福祉。

关键词：瑞典，斯堪的纳维亚半岛，优生学，种族，社会政策，生物学，北欧，种族卫生学，精神疾病

On August 20, 1997, *Dagens Nyheter,* one of Sweden's largest newspapers, published an article about the country's little-known sterilization practices involving the mentally and physically disabled, drawing harsh comparisons to the horrors of Nazi Germany.[1] The article sparked a national debate, with some claiming that the welfare state promoted sterilization laws, while others argued that social policies had eliminated them. When Maciej Zaremba wrote his article, most Swedes had never heard of the forced sterilizations that were performed within their borders, and little scholarly research had been conducted at the time. International news outlets picked up on the national debate and echoed Zaremba's comparisons to Nazi Germany. In 1997, *The Economist* stated, "The Nazis were not alone in viewing Nordic peoples as the ideal biological 'type.' A lot of Nordics, it now seems, immodestly felt the same way."[2] The *Independent* claimed "The Swedish government could face thousands of legal claims for compensation because of a Nazi-style campaign of forced sterilization of women that historians say has been hushed for years."[3] These statements promoted Nazi imagery in the public's mind, but it also captivated the interest of scholars who produced new research on the matter, providing valuable insight into how the eugenics movement had developed within Sweden.

At the turn of the twentieth century, eugenics movements emerged in Europe and the United States, as the concept of selective breeding in humans garnered public support. Between 1935 and 1975, 62,888 sterilizations were performed in Sweden, based on a law passed in 1934 and a modified version that passed in 1941. Both versions of the law permitted officials to sterilize individuals without consent in certain situations. Under the Sterilization Act of 1934, officials could sterilize those who did not have the mental capacity to understand the meaning of sterilization and its consequences. The 1941 version promoted consensual sterilization, but officials still had the means to perform forced sterilizations, without physical violence or restraint.[4] These laws mostly targeted women who had mental disabilities or hereditary diseases, and they have left a black stain on Sweden's past. Today, Sweden is renowned for its policies protecting and supporting human rights, so how could these types of laws develop in a country like Sweden? Social Darwinism was a prominent belief within the academic community and factors such as mass emigration to the United States worried scholars that a decline of the Swedish people might occur. Eugenicists relied on the support of political parties and the social elite to promote and incorporate their ideologies into social policy.

Race-based science flourished in academic circles across Europe at the beginning of the twentieth century, and eugenicists sought to define what made one race superior over another, while trying to improve gene pools. Herman Lundborg (1868–1943), a Swedish eugenicist, argued that racial purity ensured that specific traits continued to appear in subsequent generations, while new variations in the population derived from the mixing of races. If strong, racially pure individuals procreated within their own population, then they would continue the superior line, ensuring the development of the nation-state. On the other hand, he argued that populations of mixed races made society sluggish and weak, eventually leading to the death of the nation.[5] Moreover, eugenicists acknowledged that undesirable traits such as mental illness existed in their own native populations and they sought to eliminate these characteristics through social policies. Before Swedish eugenicists could pursue any form of social policy, they had to define the characteristics that made individuals superior or inferior.

Swedish eugenicists defined the perfect Swede in terms of physical and national characteristics. They argued that the Swedish race originated from ancient Germanic peoples, and their blood connection ensured they inherited numerous positive characteristics. Lundborg described the racially pure Swede as "tall and strong. The head and face were relatively long, the complexion fair and ruddy, the hair fair, and the nose most often short and straight ... light eyes."[6] Moreover, the Swedish national character portrayed Swedes as heroic, courageous, hardworking, compassionate, and hospitable, especially toward strangers. Scholars acknowledged that Swedes derived from different socioeconomic backgrounds, but diligent Swedes ensured they kept themselves well dressed and clean no matter their station. Most importantly, Swedes sought to further their education through continuous reading and used their knowledge to pursue noble endeavors. Swedes were not lazy, but hardworking, independent workers who ensured success in whatever ventures they set their minds to.[7] Even though scholars had different ideas of the perfect Swede, they used these general ideas to draw comparisons between other racial groups, defining inferior and superior characteristics within the population.

Historically, the terms "Swede" and "Swedish" described the native population of Sweden, but excluded minority groups such as the Sami, previously known as the Lapps or Laplanders, and Finns that lived within the same territory. For thousands of years, the Sami, an indigenous population, lived in the northern parts of Norway, Sweden, and Finland, and their culture centered around reindeer herding. Although they were an indigenous population of the same state, Swedes regarded them as a backwards minority that produced weakened offspring with undesirable traits. Officials wanted to keep Sami and Swedes separate and so the Sami had their own education and judicial systems. Moreover, the Swedish government forbade the Sami from settling outside certain jurisdictions and they had to remain reindeer herders as they were barred from all other occupations. To promote racially pure children, Swedes and Sami could not intermarry.[8] Lundborg described the Sami as short, dark-skinned, and unable to grow proper beards. He also noted that the Sami lived in independent communities throughout northern Scandinavia, with a meager population that never united to form a great nation.[9] If the Sami reproduced with racially pure Swedes, then their offspring would inherit undesirable characteristics, weakening the superior traits of the Swedish parent.

For hundreds of years, Finland and Sweden shared a common monarch and government, but Finns and Swedes remained two distinct groups, with Swedes considering the Finns to be inferior. Lundborg described the Finns as "short in stature, thick set and strongly built. He is fair and, as a rule, has light eyes, his hair is straight and coarse, his complexion is fair and often of a somewhat dirty grey color .... The nose is very clumsy with a concave bridge."[10] Moreover, scholars claimed that Swedes had the ability to take on leadership roles and to take on subordinate positions without any issues. However, Finnish blood made mixed Swedes defiant and undisciplined.[11] Finland and Sweden's long history together ensured that their native populations intermarried, and many of these marriages occurred in the lower socioeconomic class. Eugenicists often attacked the lower class for their laziness and other undesirable traits, blaming intermarriages for producing weakened offspring. The Finns and Sami made up most minorities living in Sweden, but Lundborg also briefly mentioned several other "alien races" that lived among the Swedes, including Jews and Romani.[12] Lundborg's prejudices against the Sami, Finns, Jews, and Romani were compatible with a variety of political viewpoints and scientific fields. Most proponents of eugenics were academics, with the main aim of protecting the biology of the Swedish population from outside genetic threats. On the other hand, Social Democrat officials argued that eugenic policies made it affordable for the welfare state to ensure lifelong security for its citizens.[13]

Moreover, scholars acknowledged that the native Swedish population had abnormal individuals that could pass on inferior characteristics to future generations. Swedish eugenicists condemned alcoholics and those infected with venereal diseases, describing them as "racial poisons." Alcoholics were not productive members of society, while sufferers of venereal diseases supposedly had numerous other hereditary diseases. Hereditary and venereal diseases were claimed to cause sterility, miscarriage, and inferior offspring. Along with mental illness, all these elements weakened the Swedish race. Eugenicists sought to improve the superior Swedish race through eugenic social policies that helped prevent inferior individuals from procreating with those deemed superior, while eliminating inferior characteristics, such as mental illness.

On November 24, 1859, Charles Darwin published his groundbreaking work *On the Origin of Species*, which introduced the concept of evolution. He argued that through natural selection, organisms that are better adapted to their environment tend to survive and produce more offspring. Subsequent generations remain well adapted and strong because the best stock of the species presumably reproduces. In 1871, Darwin published *The Descent of Man and Selection in Relation to Sex*, which applied evolutionary theory to humans, including its impact on society. Darwin argued that people developed the capacity for sympathy, especially towards the hardships of others, and a moral obligation to help those who need it, even though it may be detrimental to the helper. He wrote:

> The aid which we feel impelled to give to the helpless is mainly an incidental result of the instinct of sympathy, which was originally acquired as part of the social instincts .... Nor could we check

> our sympathy, even at the urging of hard reason, without deterioration in the noblest part of our nature. The surgeon may harden himself whilst performing an operation, for he knows that he is acting for the good of his patient; but if we were intentionally to neglect the weak and helpless, it could only be for a contingent benefit, with an overwhelming present evil. We must therefore bear the undoubtedly bad effects of the weak surviving and propagating their kind.[14]

Basically, it was a moral obligation to help those who needed it, even if it was a person of "poor quality," and helping him or her risked the possibility of him or her reproducing, lowering the quality of the next generation. Adding to the problem, the weaker and unhealthier individuals were left at home to reproduce, while the government conscripted or enlisted the most able-bodied men for their standing militaries, lowering the stock of the population. He believed that it was vital for any society to have a group of wealthy, highly educated men that had the time to dedicate to business, philosophical, or scientific endeavors because progress occurred through their efforts.[15] Darwin's argument essentially separated people into "poor" and "good" breeding stock. He emphasized the importance of limiting the "poor" breeding stock from marrying and reproducing, while encouraging the "good" stock.

Thomas Malthus (1766–1834), an English cleric and economist, published his *Essay on the Principles of Population* in 1798 and travelled to Scandinavia in the summer of 1799 to gather material for a revised edition. He believed there was a relationship between the available land and resources and the size of a population. If the population size were to exceed the amount of land and resources available, then poverty and misery would result. On his journeys through Scandinavia, he concluded that Norwegians enjoyed a better standard of living than the Swedes because their smaller population lowered the pressure on the land. He credited this to Norwegians marrying later than Swedes, which delayed children.[16] Populations could build strong, thriving societies if their size remained equal to the available land. He regarded those who could procure foods and other necessities the strongest of society, since they would survive the struggle against other people. In other words, the impoverished low class was weak, while the rich upper class was superior. These ideas were highly influential in academic circles. Scholars from around the world applied these ideas to their own countries' circumstances.

Darwin's scientific theory of evolution reached Sweden's academic circles in the late nineteenth century, and began to appear in the writings of prominent Swedish scholars such as Viktor Rydberg (1828–1895). Rydberg was a famous Swedish writer and member of the Swedish Academy, and was highly active academically and politically. He worried about the future quality of the Swedish people, believing it was in steep decline, which he ex-https://digitaltmuseum.se/021015978474/herman-lundborg

Herman Lundborg, 1920, photographed by Calla Sundbeck, Grenna Museum

VIKTOR RYDBERG

https://sok.riksarkivet.se/sbl/mobil/Artikel/6269

Viktor Rydberg, 1876, Riksarkivet

https://www.si.edu/object/charles-benedict-davenport-1866-1944:siris_arc_296652

Charles Benedict Davenport, 1866-1944, Smithsonian Institution Archives

Fisher girl from Skåne, Nordic Type, photographed by Alfred Nilson, from Herman Lundborg's *Svenska Folktyper*, 1919.

pressed in 1895 in his essay "The Future of the White Race." In this essay, Rydberg compares Europeans to Asians, expressing his fears of the decline of European populations while the supposedly inferior Asians were on the rise. One of his major worries stemmed from mass emigration to the United States. He wrote,

> In addition, European emigration also contributes to America and Australia. For what the white race loses in Europe, it gains in the colonies on the other side .... And yet here in this country (Sweden) we don't suffer from overpopulation, but from underpopulation.[17]

Population loss contributed to the worry of scholars who embraced social Darwinism, especially since it meant that there were fewer healthy able-bodied men and women to create the next generation. Sweden underwent several major blows to their population in the late nineteenth and early twentieth centuries.

Between 1850 and 1930, an estimated 1.3 million Swedes emigrated to the United States, with most settling in Illinois, Wisconsin, and Minnesota. Many of these Swedes sought to escape the hardships that plagued their homeland for golden opportunities promised across the Atlantic. For instance, between 1866 and 1868, Sweden's weather turned extremely cold, causing lakes and rivers to remain frozen until June; in the spring, crops could not be planted. Temperatures remained cold throughout the summer and autumn months, ensuring crop failures and severe food shortages throughout the country. High inflation left many people unable to afford food. The Swedish Famine killed 270,000 people, or about 15 percent of the Swedish population.[18] Many families received letters from family and friends in the United States boasting about the goodness of the new land, where "golden opportunities held out to rich and poor alike."[19] Curious about these wonderful claims, families sent their young men to visit and decide if they were true. If they decided that the United States promised a better future, then whole families would emigrate. For a small country like Sweden, these losses in population were significant. Moreover, US public health officials scrutinized immigrants at entry ports for infectious diseases, such as tuberculosis, mental impairments, and epilepsy. Officials deported those deemed unfit or too sick, forcing these individuals to return home. Scholars such as Rydberg worried that the loss of so many young healthy men and women to emigration, starvation, and disease would lower the quality of the Swedish population.

On the opposite side of the spectrum, scholars worried that massive immigration would flood the country with inferior races that would populate the country, mixing their inferior elements into the general population. Professor Johan Hultkrantz (1862–1938) claimed,

> Immigration to Sweden implies not so seldom a danger for the future of the Swedish race and above all because especially during the latest years an invasion has been taking place from the

> East, where the human material, in respect to the mental and physical qualities of the race, can hardly match itself against the ancient Swedish population.[20]

Even though Sweden remained neutral during the First World War, the influx of asylum seekers stirred fear in eugenicists. They worried that immigrants would reduce the quality of the Swedish character through inbreeding. The war helped to fuel the flames to push for eugenics-based legislation that stemmed the flow of undesirable groups of people into the country.

Officially, Sweden's last war ended August 14, 1814 with the signing of the Convention of Moss, a peace agreement between Norway and Sweden. Under the agreement, Norway and Sweden entered a union where they shared a common monarch but had separate parliaments and institutions. In 1905, the union dissolved peacefully, as Norway declared its independence. Some Swedes wanted to take up arms and force Norway back into the Union, especially since they felt that Sweden as a country lacked any form of strength or power. Ever since 1814, the Swedish government has opted for the peaceful route or neutrality. Gustaf Sundbärg (1857–1914), a Swedish statistician, was outraged with the "Swedish national character," believing it to be too timid and low in self-esteem.[21] These factors played into the idea that the Swedish society was in decline, and some scholars, such as eugenicists, and officials believed that something had to be done to improve its outlook.

In the 1860s, an Austrian scholar named Gregor Mendel (1822–1884) introduced the scientific theory of inheritance, which proposed that people inherit a single gene from each of their parents to create a pairing of two genes. Genes express themselves through a variety of characteristics, such as eye and hair color. On December 10, 1910, the Mendelian Society was founded in Lund, Sweden, with the primary purpose of stimulating hereditary research. One of its members was Herman Nilsson-Ehle (1873–1949), Sweden's first professor of genetics.[22] Nilsson-Ehle was well known for his genetic research in plants, discovering how to breed a form of wheat resistant to the winter cold. Research showed that specific breeding could strengthen and alter the targeted genes. Although Nilsson-Ehle worked primarily with plants, he was an advocate of eugenics, believing that his research could be applied to human beings.[23] Other members of the Mendelian Society were zoologists, botanists, biologists, and medical students, and they frequently discussed eugenics. The idea of eugenics spread throughout the academic community, with advocates from a variety of scientific fields.

The Swedish Society for Racial Hygiene, established in 1909, advocated Mendelian-based research, especially if it promoted eugenic social reform. The society's main goal was to influence social policy and public opinion by spreading information about eugenics and supporting genetic research. Like the Mendelian Society, it had a diverse group of scholars, with each member having his own interpretation of eugenics and political affiliation. Its members understood that biology and social reforms were above their political and scientific differences, so they sought to work together to push for eugenic social policies. Several members, such as Johan Hultkrantz, had strong left-leaning tendencies, while others, such as Herman Lundborg and Herman Nilsson-Ehle, were rightwing conservatives. Members selected Lundborg as the institute's director, which is why the society promoted rightwing eugenic policies.[24] Despite their political and scientific differences, members were deeply committed to the cause of eugenics-based social reforms. Moreover, the various academic circles worked together toward their common goal to form a type of eugenics network.

Swedish academics wanted the establishment of genetics to formally recognize eugenics as a legitimate academic disciple, and they accomplished this by establishing a position and institute for inheritance research for Nilsson-Ehle. Moreover, they strengthened the academic credentials of genetics by founding the scientific journal *Hereditas*. Societies invited scholars to lecture on racial hygiene and eugenics policies, spreading information on eugenics throughout the academic community.[25] Besides the academic community, scholars had to educate the public of the benefits of eugenics, and they realized this was a delicate task. Public opinion would have an impact on any potential social reform, so it had to sway the voting population toward pro-eugenics.

Scholars raised public awareness through exhibitions and the publications of books and pamphlets on eugenics. In 1919, the Swedish Society for Racial Hygiene hosted an exhibition held in major cities such as Stockholm, Visby, and Gothenburg, featuring photographs of people labeled as high- or low-quality. Lundborg published pamphlets that provided an overview of the exhibit. He reported that more than 40,000 people attended the exhibit and that it successfully influenced many of its visitors from various associations to support eugenics-based research and reforms. Moreover, he claimed that remote municipalities would conduct their own research on local biological features to further the field's understanding of Swedish physical characteristics.[26] Lundborg published books containing essays about different aspects of genetics and eugenics written by well-respected scholars, such as Nilsson-Ehle. In "The Struggle for Race Improvement in Sweden," Johan Hultkrantz and Emanuel Bergman argued that it was useless to improve the social environment in order to protect the race from deterioration. They favored social reform that enabled the favorable selection of parents who could transmit their superior genes to the next generation. Hultkrantz and Bergman wrote, "Thus it is of importance to seek to hinder, as far as possible, the reproduction of inferior individuals, and to increase the nativity among the better stock instead, as well as prevent immigration of inferior, and emigration of the fittest individuals."[27] The promotion of eugenics involved a wide range of topics, including marriage and immigration; forced sterilizations were a minor subsection of the larger picture. Eugenicists wanted the fittest people to have more children, but they realized that large families placed an enormous financial burden on the breadwinner. It was argued that state policy should help lighten the financial burden of large families by helping them secure a home, supplementing wages, and providing a reduction in rates and taxes. They believed that placing these types of families in a stronger financial position would help them prosper. With the promise of financial and housing security, it was hoped that family of good stock would continue to grow their families.[28] Through their promotional efforts, more scholars, government officials, and the public started to accept some of the ideas arising from eugenics.

News articles from 1997 focused on the idea that the Swedish eugenics movement was about forced sterilizations and the laws that permitted them, but this is a misconception within the general international community. Sterilization of the mentally disabled and others deemed unfit for reproduction was a topic of interest to the eugenics movement, but supporters of eugenics sought social reforms in a variety of areas, including marriage and childcare laws. In 1919, the Swedish Parliament provided a minor grant to the University of Uppsala to establish a small race-biological institution, but a group of physicians and economists wanted to create an entire school dedicated to research on race biology and eugenics. At first, Parliament disagreed with these sentiments, but in 1921, they agreed to a fully state-funded institution. In the following year, the Swedish State Institute for Race Biology was established, with Lundborg as its first director.[29] Its primary aim was to conduct research on topics related to eugenics and human genetics. Policymakers formally recognized eugenics as a legitimate field of science through the establishment of the institution, and they sought advice from eugenicists on relevant policies and social reforms. In 1910, Charles Davenport (1866–1944), an American biologist, established the Cold Spring Harbor Laboratory's Eugenic Record Office, which carried out surveys on hereditary conditions. Harry McLaughlin (1880–1943), an American sociologist, was its director and spoke in favor of implementing sterilization laws and restricting immigration of undesirable populations to Congress.[30] American politicians listened to the arguments of specialists on eugenic policies from these programs, but their connection to the state was limited. In Sweden, the Institute was the first fully state-funded school in the world dedicated to racial policy, so there was deep connection with the school and government officials.

Social policy that was based on eugenics had the basic principle of strengthening the Swedish race while weeding out elements that weakened it. Like their counterparts in the United States and Germany, Swedish academics promoted both "negative" and "positive" measures to achieve these ends. Positive measures increased reproduction among the fit elements of the population, while negative measures sought to prevent the procreation of those who were unfit.[31] An old marriage law from 1734 prevented anyone with epilepsy or any mental impairment or venereal disease from marrying, and developing a disease after marriage was enough reason to annul the marriage. In the early twentieth century, the Swedish Parliament wanted to revise the old law and sought the advice of the Institute. Doctors recommended that those with epilepsy or mental illnesses not caused by external factors or venereal diseases should not be allowed to marry. However, those with epilepsy or mental illness could receive a health certificate from a physician, proving they were fit for marriage. Doctors wanted to add restrictions on those suffering from leprosy, advanced tuberculosis, alcoholism, or physical impairments, but they felt it was pushing too much, too soon. After considering the advice from the Institute, Parliament revised the marriage law, stating that the both parties had to sign an agreement that they did not suffer from any of the mentioned ailments. If in the last three years they had suffered from them, then they had to have a doctor's certificate declaring there were no current signs of illness.[32] Although Parliament did not follow the Institution's recommendation down to the letter, this example shows how the government took eugenics into consideration in their final decision on the marriage law. They actively sought advice from eugenics experts and incorporated their ideology into law. Marriage often led to the creation of families, theoretically reducing the risk of "unfit" people from reproducing. Hultkrantz and Bergman admitted that marriage laws helped to hinder undesirable people from reproducing, but it was not foolproof. Further laws involving sterilization and segregation were vital components of increasing racial hygiene.

Eugenicists argued that Sweden's prison and asylum laws were not harsh enough and advocated that sentences for criminals and the severely mentally impaired should be lengthened. Those who committed sexual offenses should remain locked up in institutions. Separating these "worthless individuals" from the general population would lower the risk of them procreating and passing down their undesirable traits to the next generation.[33] Hultkrantz and Bergman briefly write about sterilization in their essay "The Struggle for Race Improvement in Sweden." They wrote,

> The Medical Faculty touched on this question also but considered that regulations respecting the sterilization of physical and mental degenerates by means of operating, ought not to be introduced before public opinion has been well-prepared to support it. The question has been discussed among doctors and one or another of these have, with the consent of the patient, sometimes performed an operation for sterilization on account of eugenic indications. Sterilization is a necessary complement to prohibition of marriage, and it is therefore to be hoped that the

> general public will be brought gradually, by means of continued efforts for their enlightenment to a right understanding of this weighty question.[34]

Scholars had firmly established eugenics and race biology as a legitimate field of science, especially with the Institute and politicians seeking the advice of its doctors on social policy. By the 1930s, there was a grand shift in Swedish society, where individual rights were sidelined for an efficient, productive welfare state that cared for the majority. Swedish politicians started to seek out welfare policies that benefited the state.

During the interwar period, Sweden entered into the modern era through rapid industrialization, inducing people to move from rural to urban areas. Urbanization created numerous social problems such as poor housing conditions and healthcare.[35] Hultkrantz states, "Above all the circumstances that people dwell so closely packed together, the greater extent of inebriety and sexual diseases, and, as a rule, the more unhygienic work in the towns act unfavorably on their inhabitants. A danger that is underrated is the strong mixture of blood which usually occurs in industrial centers which flourish rapidly."[36] Officials noted these poor conditions and wanted to use active welfare and population policies to improve the general population's living conditions. Sociopolitical and economic perspectives started to influence social policy, as politicians wanted to encourage productivity in the welfare state. In 1922, the Swedish Parliament argued that institutional care for the mentally handicapped placed a heavy economic strain on society. Moreover, they argued about the eugenic issue of them procreating, especially since they were deemed unfit to raise children. Children from unfit families had to be taken away and raised in proper homes to hopefully turn them into productive citizens. Under the proposed bill, the mentally impaired, epileptics, and sex offenders would be sterilized voluntarily; members of the Social Democrat, Liberal, and Agrarian parties supported the proposition.[37] Despite agreeing with the arguments and claiming it was of national interest to keep the race healthy, Parliament set up the Commission on Sterilization in 1927 to investigate the issue further, without putting the bill to a vote. Scholars continued to push the interest of eugenics through the publication of books, pamphlets, and lectures, but nothing further happened with respect to the sterilization question until the following decade.

In 1933, Parliament revisited the sterilization question, but decided to abandon the idea of voluntary sterilization. They turned their attention to the forced sterilization of individuals who were legally incompetent, or those unable to make a legal decision on their own. It was also suggested that minors could be sterilized regardless of their parents' wishes, since the needs of society were considered more important. Officials emphasized that physical violence and restraint were illegal, and it was best to prepare the mentally ill for the operation through private discussions with their doctor. Doctors were encouraged to persuade patients to accept sterilization and freely undergo the operation.[38] Under these conditions, Parliament passed the first Sterilization Act in 1934, which went into effect on January 1, 1935. When the Sterilization Act of 1941 was discussed by the Swedish Parliament, the Minister of Justice Karl Gustaf Westman claimed, "an important step in the direction of a purification of the Swedish stock, freeing it from the transmission of genetic material which would produce, in future generations, such as individuals as are undesirable among a sound a healthy people."[39] Despite politicians viewing sterilization from an economic standpoint, race rhetoric continued in the discussions. The revised bill was passed.

Sweden's sterilization program continued into the 1970s, but sterilization on eugenic grounds slowly decreased in the 1950s. One of the primary reasons for the shift in attitudes toward eugenics was due to the shift from the state to the individual. Social policies started to focus on the specific needs of the individual, such as their housing and family situations. Sterilization policies reflected the needs and desires of the state, completely neglecting the rights of the individual. When the public started to care about the rights of the individual, forced sterilization practices fell out of favor. Sterilization was a personal matter and choice and not to be used as a tool of the state for population policy.[40] Moreover, rapid economic growth in Sweden raised the standard of living for almost everyone, granting individuals greater autonomy and freedom. Finally, the general level of education improved, permitting the average person to speak out in debates and discussions, so their opinions and experiences were heard. Average citizens took part in political discourse, rather than only politicians and academics.[41] These social shifts reinforced the rights of the individual above the wellbeing of the state.

After Charles Darwin published his famous works on natural selection and its application to human beings, the eugenics movement started to take root, as more and more scholars worried about racial hygiene. Academics with different scientific and political backgrounds worked together towards the common goal of eugenics reform by educating the public on the benefits of population policies. Through their continuous promotional efforts, eugenics became an established and legitimate science in the field of genetics. Government officials accepted and supported racial doctrine, fully funding the first race biological institution in the world. They based their ideas on the core principle that fit persons should be encouraged to procreate, while lowering the risk of the unfit reproducing through policies such as sterilization and segregation. Parliament regularly sought the advice of its doctors on social policies, implementing some of their recommendations into law.

By the 1930s, the needs of the welfare state took precedence over individual rights, as Sweden sought to improve social conditions, such as housing and healthcare, through policies that benefited society, regardless of their impact on specific individuals. Thousands of forced sterilizations were carried out on the mentally ill for eugenic reasons, but these numbers started to decline as public sentiments reversed. Individual rights became more important than the wellbeing of the state. Eugenics-based policies impacted a variety of social policies, including marriage, childcare, and immigration, and this could not have occurred without the tireless efforts of scholars campaigning to teach the public about its supposed benefits.

Bibliography

Bergfors, Georg. "The Swedish National Character." In *The Swedish Nation in Word and Picture.* Edited by Herman Lundborg and John Runnström, 34–38. Stockholm: Hasse W. Tullberg Co., 1921.

Bjorkman, Maria and Sven Widmalm. "Selling Eugenics: The Case of Sweden." *Notes and Records of the Royal Society* 64 (2010): 379–400. https://www.jstor.org/stable/25802126.

Broberg, Gunnar and Mattias Tyden. "Eugenics in Sweden: Efficient Care." In *Eugenics and the Welfare State.* Edited by Gunnar Broberg and Nils Roll-Hansen, 77–149. East Lansing: Michigan State University Press, 2005.

Buchanan, Allen, Dan Brock, Norman Daniels, and Daniel Wikler. *From Chance to Choice: Genetics and Justice.* Cambridge: Cambridge University Press, 2000.

Collins, Francis. "Foreword: This Past Must Not Prologue." In *Medicine After the Holocaust.* Edited by Sheldon Rubenfeld, xix–xx. New York: Palgrave MacMillan, 2010.

Crandall, Rosa. *The Neil Family, Sweden-America, 1718–1908.* London: A.M. Eddy Press, 1908.

Darwin, Charles. *The Descent of Man and Selection in Relation to Sex,* 2nd ed. London, UK: John Murray, Albemarle Street, 1890.

Dianda, Bas. *Political Routes to Starvation: Why Does Famine Kill?* Wilmington, DE: Vernon Press, 2019.

Hoglund, Mattias and Bengt Bentsson. "The Origin of the Mendelian Society in Lund and the State of Hereditas." *Hereditas* 151 (2014): 110–14. https://onlinelibrary.wiley.com/doi/pdf/10.1111/hrd2.00078.

Hultkrantz, Johan and Emanuel Bergman. "The Struggle for Race Improvement in Sweden." In *The Swedish Nation in Word and Picture.* Edited by Herman Lundborg and John Runnström, 71–80. Stockholm: Hasse W. Tullberg Co., 1921.

Lundborg, Herman. "The More Important Racial Elements That Form a Part of the Present Swedish Nation." In *The Swedish Nation in Word and Picture.* Edited by Herman Lundborg and John Runnström, 23–33. Stockholm: Hasse W. Tullberg Co., 1921.

Lundborg, Herman. *Svenska Folktyper.* Translated by Susan Danielsson. Stockholm: AB Hasse W. Tullbergs, 1919. https://coffe.files.wordpress.com/2012/07/svenska_folktyper.pdf.

Lundborg, Herman and Frans Linders. *The Racial Characters of the Swedish Nation.* Stockholm: Almqvist Wiksell Printing Company, 1926.

Magnusson, Lars. *An Economic History of Sweden.* London, UK: Routledge, 2000.

Mattson, Greggor. "Nation-State Science: Lappology and Sweden's Ethnoracial Purity." *Comparative Studies in Society and History* 56, no. 2 (April, 2014): 320–60. https://www.jstor.org/stable/43908505.

"Nordic Eugenics – Here, of all places." *The Economist.* August 28, 1997. https://www.economist.com/europe/1997/08/28/here-of-all-places.

Rydberg, Viktor. "Den Hvita Rasens Framtid." http://runeberg.org/rydvaria/hvita.html.

Spektorowski, Alberto. "The Eugenic Temptation in Socialism: Sweden, Germany, and the Soviet Union." *Comparative Studies in Society and History* 46, no. 1 (Jan., 2004): 84-106. https://www.jstor.org/stable/3879414.

"Steriliseringsfrågan i Sverige 1935–1975 Historisk belysning - Kartläggning – Intervjuer." *Regeringskansliet.* March 1, 2000. https://www.regeringen.se/rattsliga-dokument/statens-offentliga-utredningar/2000/03/sou-200020/.

"Sweden admits to racial purification." *Independent.* August 25, 1997. https://www.independent.co.uk/news/world/sweden-admits-to-racial-purification-1247261.html.

Zaremba, Maciej. "Rasren i Välfärden. Folkhemmets Förträngda arv." *Dagens Nyheter.* August 20, 1997. https://www.dn.se/arkiv/kultur/rasren-i-valfarden-folkhemmets-fortrangda-arv/.

Notes

1 Maciej Zaremba, "Rasren i Välfärden. Folkhemmets Förträngda arv," *Dagens Nyheter*, August 20, 1997, https://www.dn.se/arkiv/kultur/rasren-i-valfarden-folkhemmets-fortrangda-arv/.

2 "Nordic Eugenics – Here, of All Places," *The Economist,* August 28, 1997, https://www.economist.com/europe/1997/08/28/here-of-all-places.

3 "Sweden Admits to Racial Purification," *Independent*, August 25, 1997, https://www.independent.co.uk/news/world/sweden-admits-to-racial-purification-1247261.html.

4 "Steriliseringsfrågan i Sverige 1935–1975 Historisk belysning - Kartläggning – Intervjuer," *Regeringskansliet,* March 1, 2000, https://www.regeringen.se/rattsliga-dokument/statens-offentliga-utredningar/2000/03/sou-200020/

5 Herman Lundborg, "The More Important Racial Elements That Form a Part of the Present Swedish Nation," in *The Swedish Nation in Word and Picture,* ed. Herman Lundborg and John Runnström (Stockholm: Hasse W. Tullberg Co., 1921), 25.

6 Ibid., 27.

7 Georg Bergfors, "The Swedish National Character," in *The Swedish Nation in Word and Picture,* ed. Herman Lundborg and John Runnström (Stockholm: Hasse W. Tullberg Co., 1921), 36–37.

8 Greggor Mattson, "Nation-State Science: Lappology and Sweden's Ethnoracial Purity," *Comparative Studies in Society and History* 56, no. 2 (April, 2014): 337, https://www.jstor.org/stable/43908505.

9 Lundborg, "The More Important Racial Elements," 31.

10 Ibid.

11 Georg Bergfors, "The Swedish National Character," 35.

12 Lundborg, "The More Important Racial Elements," 33.

13 Allen Buchanan, Dan Brock, Norman Daniels, and Daniel Wikler, *From Chance to Choice: Genetics and Justice* (Cambridge, MA: Cambridge University Press, 2000), 32.

14 Charles Darwin, *The Descent of Man and Selection in Relation to Sex,* 2nd ed. (London, UK: John Marray, Albemarle Street, 1890), 134.

15 Ibid., 135.

16 Lars Magnusson, *An Economic History of Sweden* (London, UK: Routledge, 2000), 15-16.

17 Viktor Rydberg, "Den Hvita Rasens Framtid," accessed January 8, 2020, http://runeberg.org/rydvaria/hvita.html.

18 Bas Dianda, *Political Routes to Starvation: Why Does Famine Kill?* (Wilmington, DE: Vernon Press, 2019), 271.

19 Rosa Crandall, *The Neil Family, Sweden-America, 1718–1908* (London, UK: A.M. Eddy Press, 1908), 8.

20 Johan Hultkrantz and Emanuel Bergman, "The Struggle for Race Improvement in Sweden," in *The Swedish Nation in Word and Picture,* ed. Herman Lundborg and John Runnström (Stockholm: Hasse W. Tullberg Co., 1921), 74.

21 Gunnar Broberg and Mattias Tyden, "Eugenics in Sweden: Efficient Care," in *Eugenics and the Welfare State,* ed. Gunnar Broberg and Nils Roll-Hansen (East Lansing: Michigan State University Press, 2005), 78.

22 Mattias Hoglund and Bengt Bentsson, "The Origin of the Mendelian Society in Lund and the State of Hereditas," *Hereditas* 151 (2014): 111–12, https://onlinelibrary.wiley.com/doi/pdf/10.1111/hrd2.00078.

23 Maria Bjorkman and Sven Widmalm, "Selling Eugenics: The Case of Sweden," *Notes and Records of the Royal Society* 64 (2010): 380, https://www.jstor.org/stable/25802126.

24 Ibid., 382.

25 Ibid.

26 Herman Lundborg, *Svenska Folktyper,* trans. Susan Danielsson (Stockholm: AB Hasse W. Tullbergs, 1919), 1, https://coffe.files.wordpress.com/2012/07/svenska_folktyper.pdf.

27 Hultkrantz and Bergman, "The Struggle for Race Improvement in Sweden," 71.

28 Ibid., 76.

29 Herman Lundborg and Frans Linders, *The Racial Characters of the Swedish Nation* (Stockholm: Almqvist Wiksell Printing Company, 1926), 7–8.

30 Francis Collins, "Foreword: This Past Must Not Prologue," in *Medicine After the Holocaust,* ed. Sheldon Rubenfeld (New York: Palgrave MacMillan, 2010), xix.

31 Bjorkman and Widmalm, "Selling Eugenics," 381–82.

32 Hultkrantz and Bergman, 73.

33 Ibid., 74.

34 Ibid.

35 Alberto Spektorowski, "The Eugenic Temptation in Socialism: Sweden, Germany, and the Soviet Union," *Comparative Studies in Society and History* 46, no. 1 (Jan., 2004): 93, https://www.jstor.org/stable/3879414.

36 Hultkrantz and Bergman, 75.

37 Broberg and Tyden, "Eugenics in Sweden: Efficient Care," 101.

38 Ibid., 102.

39 Ibid., 107.

40 Ibid., 134.

41 Ibid.

Australian and American Relations in the Southwest Pacific Theater of World War II

Alisha Hamel

American Military University

Abstract

While both the Australians and Americans fought the Japanese during World War II, leadership and cultural differences became apparent when they fought together in New Guinea. While Australia and the United States were and still are great allies, even the best of allies have different cultures, training, and leadership methods, often resulting in difficulties when they are put into combat roles together.

Keywords: WWII, Australia, Southwest Pacific Theater, MacArthur, Curtin, Blamey

Relaciones australianas y americanas en el teatro del Pacífico sudoccidental de la Segunda Guerra Mundial

Resumen

Mientras tanto los australianos como los estadounidenses lucharon contra los japoneses durante la Segunda Guerra Mundial, el liderazgo y las diferencias culturales se hicieron evidentes cuando lucharon juntos a través de Nueva Guinea. Australia y los Estados Unidos fueron y siguen siendo grandes aliados, pero incluso los mejores aliados tienen diferentes culturas, entrenamiento y métodos de liderazgo que a menudo resultan en dificultades cuando se los pone en roles de combate entrelazados.

Palabras clave: Segunda Guerra Mundial, Australia, Teatro del Pacífico Sudoccidental, MacArthur, Curtin, Blamey

第二次世界大战西南太平洋战场中的澳美关系

摘要

尽管澳洲人与美国人在二战期间共同对抗日本人，但各自的领导与文化差异在新几内亚战区变得明显。澳洲与美国从过去到现在一直是亲密联盟，但即使是最好的联盟也有不同的文化、训练和领导方法，导致他们在投入相互交织的战斗模式时出现困难。

关键词：二战，澳洲，西南太平洋战区，麦克阿瑟，柯廷，布莱梅

Australians and Americans fought together and separately in the Southwest Pacific Theater during World War II. The Australian military effort was vital to maintaining the Allied presence in the Pacific Theater, giving time for the Americans to mobilize their troops and supply chain. From 1942 to 1943, because of their readiness and location, the Australians fought in the Southwest Pacific Theater's primary offensive actions through the Allied victory on the Huon Peninsula, a strategically located peninsula in New Guinea that, along with New Britain, controlled the straits between the Bismarck and Solomon Seas.

In 1944, the Americans took over almost all primary offensive operations, while the Australian role became one of clean up operations, a mild sounding role for a very dangerous mission. This article explores the reasons that the Australians were pulled from offensive operations and relegated to secondary duties after their offensive win on the Huon Peninsula. With no mention of the Australians or any other ally in the public press, a common misperception was that the Americans had single-handedly won the Pacific War.

Australian and American relations during World War II were built on a common background and a mutual foe, which resulted in them becoming strong allies. Australia saw the advantages of cultivating strong relations with the United States both to create an international voice and to protect themselves from a Japanese attack. The United States needed Australia as a large staging base in the Pacific Theater and saw that Australia also had extensive natural resources, had already built civil and military facilities, and an experienced, though small, military.

The beginning of the war in the Pacific Theater required close cooperation between Australia and the United States, but even close allies have different ideas about how to fight a war and how to use their troops. Differing styles

General Sir Thomas Blamey. March 26, 1945. Bougainville, Solomon Islands Australian War Memorial 079976

of military leadership and training had to be synchronized, supply issues had to be resolved, and even the national tastes for distinct types of food to be consumed had to be addressed.

Australia, with a population of almost 7 million and an army of close to 730,000 during World War II, was considered a minor Allied partner with a minor voice in Allied politics. The major powers of Russia with a population of over 170 million, Great Britain with a population of almost 48 million, and the United States with a population of over 131 million were the big three Allies of World War II. Australia went to war in 1939 as a Commonwealth country when Great Britain declared war against Germany, contributing troops to fight under English command in North Africa and the Middle East with very little representation for their use.[1]

The looming threat of war with Japan caused Australia to recognize the need for representation in the Allied councils of war in order to safeguard its own national interests. The need for representation was a difficult concept for a country that had never had to fight to have its voice heard, that had relied on British diplomatic and intelligence services, and that had little trust in its military leaders. In addition to its relatively small population, Australia had a relatively weak industrial capacity, which contributed to the difficulty of pursuing an independent strategic and foreign policy. Australia found that it was only able to influence allied strategy by using its military as a reward for the United States or Great Britain to recognize Australia's concerns.[2]

After the Japanese attack on Pearl Harbor and the Philippines in December 1941, the people of Australia and Prime Minister John Curtin could clearly see that their homeland was the next target for the Japanese due to its location and the ability to base Allied troops there for counter-offensives. At the Arcadia Conference in Washington, DC, the first in a series of high-level conferences between the United States and Great Britain (December 24, 1941–January 14, 1942), the short-lived Australian-British-Dutch-American (ABDA) Command and a Combined Chiefs of Staff (CCS) were created. One of the attendees, Australia's Minister to the United States, Richard Casey, sent a cablegram to Curtin stating; "I have reason to believe that the President [Roosevelt] will try very hard to have an American accepted as Commander-in-Chief in the Pacific and the Far East Theatre and that General MacArthur [now in the Philippines] will probably be the individual nominated. I understand that although not devoid of human frailties, he is a good man." He continued, "It occurs to me as not impossible that the headquarters of the Commander-in-Chief of the Pacific and the Far East might be in Australia."[3] Curtin asked for clarification of Australia's role with the United States and Britain, and the United States responded on January 8, 1942 with a proposal to create the Australia/New Zealand (ANZAC) area. This assured Australia that the US would help with its security and provided the United States with a staging area for American troops to train before going into battle with the Japanese.[4]

With Australia available as a staging base for US troops, Australia became more important to the Allied strategy. It was imperative to develop a strategy to maintain safe shipping lanes from the United States to Australia. Australia found a stronger voice in the Allied councils of war, which increased even more when MacArthur arrived in Australia. MacArthur had a more direct connection to President Franklin Roosevelt and General George Marshall, the Chief of Staff of the United States Army, and to the Allied Combined Chiefs of Staff. Australia's External Affairs Minister H.V. Evatt noted that MacArthur would be able to attract American troops and supplies to Australia and that this quality was even more important than his generalship.[5]

MacArthur arrived in Australia to a hero's welcome in early 1942. The Australians expected him to bring American assets into Australia to protect Australia from the Japanese, as

Australia was to be used as the staging base to develop a plan of attack to defeat the Japanese. Before MacArthur's arrival in Australia, Curtin said, "Without any inhibitions of any kind, I make it quite clear that Australia looks to America, free of any pangs as to our traditional links with the United Kingdom."[6] At that time, Australia felt like it needed American military leadership, and MacArthur was happy to provide it. MacArthur almost completely directed the Australian war effort. The Australian public adored him and treated him like a movie star. MacArthur and Curtin developed a very close and mutually beneficial relationship, and MacArthur became the de facto field marshal in charge of all of Australia's defense forces.[7] Although the Australian public loved MacArthur throughout the war, the Australian troops developed second thoughts as the war progressed.[8]

MacArthur saw the value of Australia as an island bastion from which to mount an offensive against the Japanese, and the resident Australian forces, especially in 1942 and 1943, defended and then led the offensive against the Japanese. He also recognized the importance of Australian public support. Curtin recognized the value of MacArthur, and they worked well together politically to maintain the safety of Australia and to push for an Allied attack in the Southwest Pacific Theater.

As Prime Minister, Curtin led Australia successfully through World War II and after the attack by the Japanese on Darwin in February 1942. He rejected the British strategy for Australian troops. "Australia's intention was to carry the fight to the Japanese just as soon as forces became available to do it. At this stage, however, only the first troops of the 7th Australian Division had arrived in Australia."[9] Because the 9th Australian Division continued to be needed by the British Empire to fight in the Middle East, a second US division was offered to be sent to Australia in addition to the already offered US 41st Division, a Northwest National Guard infantry division considered the "best" National Guard infantry division in the United States. The US 32nd Division, a National Guard infantry division from Michigan and Wisconsin, was pulled from training to fight in the European Theater and was sent to Australia, arriving only a month after the 41st Division arrived. Even with these two US infantry divisions and the returning Australian troops from North Africa, MacArthur still was short of fighting men and needed Australia to institute conscription. Australia had a policy against conscription. Although Curtin had been a formidable opponent of conscription, when Australians were needed to protect their homeland, he reversed course and implemented conscription and those conscripted troops helped to lead the fight in New Guinea.

Although MacArthur oversaw all of Australia's defense forces, he also commanded all Allied troops in the Southwest Pacific Theater and as such was directed to create an Allied staff. The Southwest Pacific Theater consisted of a geographical area that included the Philippines, Borneo, the Dutch East Indies, East Timor, Australia, New Guinea and Papua New Guinea, and the western part of the Solomon Islands. The Allied staff was intended to include officers from Australia, the Netherlands, the Philippines, and Great Britain, but MacArthur assigned only long-time US staff officers who had evacuated with him from the Philippines to key staff positions, resulting in a US-centric headquarters. When MacArthur became the Supreme Commander of Allied Forces in the Southwest Pacific Area in April 1942, US Army Chief of Staff General George Marshall directed MacArthur to appoint an Australian as Commander, Allied Land Forces. MacArthur chose General Sir Thomas Blamey. Blamey had a storied background. In 1936 as Victoria's commissioner of police, he had a reputation for being confrontational, violent, and ruthless. He was also well known for his public drinking and womanizing. Despite that reputation, Blamey had a distinguished military career starting with a posting to the 1st Australian Division in Egypt in World War I. He landed at Gallipoli in 1915, returned to Egypt to form the 2nd Australian Division, and ended World War I as the Chief of Staff of the Australian Corps. When World War II started, Blamey took command of the 6th Australian Division and by December 1941, had been promoted to General.[10]

Blamey commanded the Australian troops in North Africa and Greece, but when Australia was attacked by the Japanese at Darwin in 1942, he was rushed back to Australia to activate the defense, arriving just five days after MacArthur arrived from the Philippines. Blamey had major problems he needed to address immediately. He had to establish a coalition relationship with the US armed forces, create a plan for the defense of Australia, and dramatically increase Australia's insufficient armed forces. Blamey proved to be the best possible Australian to be put in that position. He stayed loyal to MacArthur until the end of the war, but politely and firmly held fast to Australian autonomy.[11]

In addition, Blamey was the only Australian military commander who could offer advice to Curtin because of his extensive combat record. The Prime Minister had very little military expertise and only trusted military leaders with proven records of accomplishment. While MacArthur had the lead in developing Australian war strategy, Blamey influenced that strategy because he also had direct access to Curtin. However, because Blamey wore two hats as the Commander, Allied Land Forces and the Commander in Chief of the Australian Militia Forces, he had limited time to influence the Prime Minister. Luckily, for the first eighteen months of the Pacific war, MacArthur had the same strategy as Australia. With the competent and strategic decisions that MacArthur was making, Australian political leadership did not have to make tough decisions, and Australia did not develop a strategic view outside of MacArthur's strategy.[12]

Blamey was a good counterpoint to MacArthur. He was loyal to MacArthur, but he had strong views on Australian sovereignty and understood the role of politics in war. Blamey fought to maintain independent use of Australia's troops with a real concern for their welfare. Without his untiring promotion of Australia's sovereignty, MacArthur would have most likely disregarded Australia's concerns.[13]

Because the composition of MacArthur's senior staff was composed of his most trusted advisors, the "Bataan Gang," this all-American staff, with the exception of Blamey, caused continued friction throughout the war with senior-level Australian commanders. There was little the Australians could do about the situation. They needed the Americans to protect Australia and force the Japanese back. The lack of respect for Australian military leadership generated bad blood between MacArthur and senior-level Australian commanders. MacArthur defended the lack of Australian representation on his staff by saying that there were few qualified senior staff officers in Australia. The Australians proved themselves in battle and fought well in the foxholes with the Americans, but there continued to be friction at the higher levels of command.[14]

The first real land defeat suffered by the Japanese happened at Milne Bay, showing that the Japanese were no longer invincible. Milne Bay was important to the Allies because of its location for the proposed attack on Rabaul and land-based support for sea movement to the northern side of New Guinea. "The first major battle of WWII in which Japanese ground forces were defeated was not Guadalcanal, but Milne Bay, Papua New Guinea. Laurels for the victory go not to the U.S. forces, but the 7th Brigade of Australian Militia and the 18th Brigade of the Australian Imperial Force."[15] These units, a militia infantry brigade and a veteran Australian Imperial Force (AIF) brigade supported by Kittyhawk fighters, stationed at Milne Bay held off a Japanese attack that lasted from August 25 to September 7, 1942. The next major battle in New Guinea for the Australians was fought over the Owen Stanley Mountains via the Kokoda Trail and into Buna and Gona.

There is no more terrible traverse to battle than over the Owen Stanley Mountains on the Kokoda Trail. "The Diggers marched on, weary, rain-soaked, but eager to get to grips with the enemy. One day was much like another—the sheer physical agony of the track, the changeless jungle scenery broken by the little clearings as village after village was retaken."[16] The Kokoda Trail at times required hand-over-hand traverse; the environment was constantly wet and became very cold as the soldiers went over the top of the Owen Stanley Mountains. Only what could be carried could be brought with them, and they were expected to fight not only along the trail, but also when they reached the other side. The men came off the trail starving, clothes in tatters, but with a fire to defeat the Japanese.

In 1943, at Buna and Gona, the Australians and Americans fought near and with each other for the first time in offensive operations in significant numbers, with the fight consisting of primarily Australians due to their location and ability to mobilize enough troops to start the fight. "Indeed throughout 1942 and 1943, the Australians bore the brunt of Southwest Pacific Area (SWPA) war. Australian troops [which had been mobilized as part of the British Commonwealth in 1939] rushed back home from North Africa or hastily deployed from Down Under, blunted the initial Japanese thrust at Port Moresby in 1942, and in the following year they spearheaded MacArthur's drive through Papua and Huon peninsula."[17]

The Americans fought for the first time when the US 32nd Infantry Division fought in September 1942 at Buna and Gona, and quickly found themselves fighting not only against the Japanese but also against the elements. At Buna and Gona, the US 32nd Infantry Division sustained almost 100 percent casualties counting those killed in action, wounded in action, missing in action, and downed by sickness due to the tropical climate. The 32nd lost its commander, Major General Forest Harding, when he was relieved of command. General Douglas MacArthur, Commander Southwest Pacific, felt that Harding had not pushed the 32nd hard enough to defeat the Japanese. The men felt that "Forest Harding was too principled to add ... 'another bloody repulse' to history's long roll of military disasters by sacrificing his soldiers on the altar of Douglas MacArthur's impatience." Harding was replaced by the I Corps commander, Lieutenant General Robert L. Eichelberger.[18]

When Eichelberger arrived at Buna and Gona to take over the battle command of the 32nd Infantry Division, he pulled all the men back from fighting, fed them a hot meal, and reassessed the situation. They were sent back into battle fighting alongside the Australian troops. After that respite, both the Americans and Australians proceeded to defeat the Japanese at Buna and Gona. "[MacArthur] had been critical of Australian performance during the Port Moresby battles, claiming that they were not good in the field or the jungle, that they were all recruited from the slums of Australia, and they lacked fighting spirit."[19] At Buna and Gona, MacArthur expected more from the US troops than he got. The US troops reacted as the Australians had reacted when they first went into battle with overconfident commanders and troops that were lax about security and camouflage.

The US 41st Infantry Division sent its first unit into combat, the 163rd Regiment, to finish off the battle at Buna and Gona under the command of the Australians, but "after Buna American combat units came under direct Australian command on only one other occasion."[20] The one other occasion was at the landing at Nassau Bay and battle for Salamaua when the 162nd Regiment of the 41st Infantry Division fought under the triple command of the Australian 17th Brigade, the 3rd Australian Division, and the US 41st Infantry Division, creating multiple command issues and giving MacArthur a reason to divide up the missions of the Americans and the Australians to ensure US dominance in offensive operations.[21]

MacArthur was intent on ensuring that the United States received the credit for winning the war in the Pacific. At first he had reluctantly relied on the Australian military, but as more US forces arrived in Australia and trained to fight in the jungles of New Guinea, MacArthur shifted the burden of the offensive to US troops. The last joint Australian and US battle happened at Salamaua (June to September 1943) when the American 162nd Regiment fought under the command of the 17th Australian Brigade.[22]

The Allied Ground Force Commander Blamey, the token Australian on MacArthur's staff, was expected to command of all of the Allied ground forces as his title implied, but MacArthur circumvented that by routinely creating task forces outside of Blamey's command. While MacArthur had publicly approved of the appointment of General Blamey as the Southwest Pacific's ground forces commander, privately he started assigning all US troops to General Walter Krueger's 6th Army, named Alamo Force, which left only Australian troops under General Blamey's authority.[23] On Saturday, February 20, 1943, *The Mercury* announced:

> An official spokesman at Gen MacArthur's headquarters, referring today to the appointment of Lt-Gen Walter Krueger to the command of the 6th United States Army in the South-West Pacific area, said it had not special operational or strategic significance and was merely a move to make administration more flexible. The 6th Army was to be made up entirely of the American troops already in the South-West Pacific area. The High Command in the area would remain unchanged, with Gen MacArthur in supreme command and Gen Sir Thomas Blamey and Lt-Gen Kenney in command of Allied land and air forces respectively.[24]

Curtin valued MacArthur's political connections and military expertise over his own military commanders, which over time resulted in the Australian Army receiving unpleasant assignments with little to no recognition or US support. Curtin's admiration worked in Australia's interest until MacArthur no longer needed Curtin as more US troops became available in the Southwest Pacific Theater and supplies from the United States increased.

MacArthur soon found that if he underreported the number and quality of Australian troops, he received more US troops. Fredrick Shedden, Australian War Council, wrote, "General MacArthur said it presented a misleading and injurious picture to Washington to talk about the total number of men in the Army in Australia, as only the two A.I.F. divisions could be considered first-class shock troops."[25] As the war progressed and more US troops arrived to fight in the Southwest Pacific Theater, MacArthur started to diminish Australian involvement in the war. MacArthur believed that since the Americans had the resources to be able to defeat Japan, the Americans should determine the strategy and claim the victory. MacArthur downgraded the Australian role subtly by referring to the battle forces as Allied Forces when Australians commanded the operations and US Forces when Americans commanded operations.[26]

MacArthur's first offensive strategic action in the Southwest Pacific Theater was to make Port Moresby the fulcrum of both the defense of Australia and the attack through New Guinea. MacArthur felt that if the Japanese got a foothold in Australia, they might prove to be unstoppable, and it was of utmost importance to protect Australia from New Guinea. MacArthur took credit for the idea that Australia had to be saved in Papua even though the Australians had already held defensive positions there ever since the Japanese had started their offensive moves into New Guinea. The Allied Air Commander for the Southwest Pacific, George Kenney, said, "MacArthur without fear of criticism might have decided to remain on the defensive until sufficient forces could be made available ... a lesser general might even have considered the abandonment of Port Moresby, his only base in New Guinea."[27] Partly because of the offensive tactic used by MacArthur and because of the Battle of the Coral Sea, the Japanese started pulling back their troops in New Guinea and failed in their attempt to take Port Moresby.

The Australians were placed in an untenable position when MacArthur consistently announced the success of each of his conquests in his drive back to the Philippines at the first sign of capitulation by the defenders even if it were to take months of fighting, primarily by the Australians, to secure the location. [28]

MacArthur, a master of the use of the press to inform the public, was not one to share the limelight with anyone, including his staff, troops, or especially the Australians. The Australians who had primarily fought and shed blood in these first offensive encounters with the Japanese were not given credit due to them. After the Battles of Salamaua and Lae and the capture of Madang, the Australians were given low-profile missions of garrison duty and were ordered to clean up the areas in New Guinea already considered conquered. This did not sit well with the Australian government. As Australian historian John Robertson wrote, "Without Australian political, logistic and military support it is hard to see how MacArthur could have made this grand return [to the Philippines], but no Australian land or air-force unit, and no Australian notables, were there [in Leyte] to share the glory."[29]

In November 1943, at the Sextant Conference in Cairo, the US Navy gave priority in the Pacific to Admiral Chester Nimitz, who was based in Hawaii, with the understanding that taking small islands would be a faster way to move toward Japan. General MacArthur had other ideas. He argued that an attack based in Hawaii would lose momentum, as the Navy would have to steam back to Hawaii after every operation to regroup before starting the next operation. The Japanese would then be able to reinforce the Mandates, German colonial islands that had been governed by the Japanese since the end of World War I as part of the Japanese colonial empire, since there was no way for air coverage while the Navy was regrouping.

MacArthur also argued that the US Navy was too wedded to the old Orange Plan scenario, one of the color war contingency plans developed by the United States before World War II. The Orange Plan was the color assigned to the plan dealing with a possible war against Japan and assumed that the United States would fight against Japan alone. After many revisions, the Orange Plan had evolved into an offensive plan, primarily naval, to take small islands on the way to rescuing the Philippines. The Philippine garrison was to hold Manila Bay until this superior naval force arrived. It was shelved in 1937 to the chagrin of the Navy, due to the near-impossibility of the Navy to be able to arrive in the Philippines in a timely manner, and replaced with the more reasonable, new natural strategic line of Alaska-Oahu-Panama. In 1939, a new series of plans called the Rainbow Plans were created. Rainbow 5 became the basis for the US strategy during World War II, calling for an alliance with Britain and France and assuming a two-front war in the Atlantic and Pacific.[30]

To MacArthur, it made more sense to move along the New Guinea coastline initially with air support from the Australian mainland and to create a series of new forward headquarters with airfields and seaports as the Allies pushed the Japanese back. He felt that the Cartwheel Campaign, an operation to neutralize the Japanese base at Rabaul using his troops to advance along the New Guinea coastline as Nimitz's forces advanced through the Solomon Islands towards Bougainville, had shown dramatic progress and presented an excellent opportunity to reenter the Philippines. MacArthur always saw the Philippines as key to defeating the Japanese.[31]

To prove his theory, MacArthur had his staff create a set of plans codenamed Reno, with the expectation that his troops would be able to reenter the Philippines at Mindanao in February 1945. The Joint Chiefs of Staff supported conquering New Guinea but were ambivalent about returning via the Philippines, and they were loath to place all of the US troops in the Pacific under MacArthur's command. MacArthur then rallied the US and Australian public with the phrase that became famous from a speech he made after his evacuation from the Philippines in 1941: "I shall return." The US public was viscerally drawn into liberating captured US troops, US civilian captives, including women and children, and the people of the Philippines.

Admiral Chester Nimitz supported MacArthur in this plan to attack along the New Guinea–Mindanao axis, but argued for a two-pronged approach. The Joint Chiefs of Staff issued the following directive: Nimitz was to take the Marianas and hit the Palaus by bypassing Truk from the north. MacArthur was to overrun New Guinea and occupy Mindanao. Nimitz and MacArthur were then expected to work together in defeating the Japanese.[32]

This strategy went against Clausewitz's principle of war, mass—a concentration of force in a single offensive push—but the two-pronged approach and dilution of resources to support each approach in this case worked especially well. The Japanese became very confused, and could not predict the location of the next Allied offensive, causing them to constantly reposition their troops. The Joint Chiefs of Staff from Washington coordinated the two prongs of command with Nimitz and MacArthur, putting up with MacArthur's posturing and bellyaching and the occasional complaints from Nimitz, and getting the best out of both commands. The Japanese, never able to concentrate on either front, whipsawed repeatedly between the two prongs of attack, losing momentum, men, and supplies with each move.[33]

Once the decision was made to focus on both prongs of the attack, Australia became a minor player in the war in the Pacific. Australia was also forced to draw down its level of combat participation because of overwhelming manpower demands at home. Australians still fought to have a seat at the table once the Japanese were defeated, but it grew harder and harder for the Australians to influence the war in the Southwest Pacific after the Americans arrived in force, set up routes for US resupply, and moved the Southwest Pacific headquarters out of Australia.[34] The Australians knew that MacArthur was downgrading their role in the Southwest Pacific, but there was little they could do about it.

By 1944, Australia had been at war for over four years, fighting in Britain's campaigns in North Africa, Greece, Crete, and the Middle East and then in Papua and New Guinea in 1942 and 1943. They had committed so many men to fight that they nearly crippled their economy and had to demobilize men to work in the fields and factories back home. The War Commitments Committee estimated a need for 78,602 men in high priority industries by the end of 1944. To solve this major shortage, Curtin directed that the Australian Army release 30,000 men and the RAAF release 15,000 men. Of these numbers, 20,000 were to be released by December 31, 1944 and the rest by June 30, 1945.[35]

Blamey's remaining Australian forces performed clean up operations and cleared the Japanese out of New Guinea as the Americans swept into the Philippines without the Australians.[36] MacArthur made it clear to Blamey that after the encirclement of Rabaul, Blamey's New Guinea Force would be assigned mopping up operations and garrison duties in increasingly rear areas. The last major Australian-led offensive happened in late April 1944, when the Australian 7th Division captured Madang just northwest of Finschafen, as MacArthur had the US forces take Hollandia, then moving the Southwest Pacific Theater headquarters to Hollandia shortly afterward.[37]

Blamey cautioned his commanders to not risk lives unnecessarily with overly aggressive tactics, but to keep up the pressure enough to not allow the Japanese to counterattack. MacArthur pressured Blamey to stay on the offensive in these "mopping up" operations, although by this time, many of the Australian Armed forces had had enough of fighting for MacArthur and preferred to fight these relatively unimportant battles in New Guinea rather than to return to MacArthur's control.[38] The Australian press started to note the unequal use of their troops, and in April 1945, the *Pacific Islands Monthly* wrote an article about the Australian troops.

> Everyone assumed that the hard-fighting Australian divisions, which had done so well beside the Americans when New Guinea represented the front line, would move on north-westwards with General MacArthur. Instead, we heard nothing whatever about them, for many months, until it was announced, late in 1944, that they were do the mopping-up they have been, ever since. They have fought tenaciously, and with their usual initiative and courage; but the people of Australia are awakening to the fact that this is a wasteful, uninspiring and depressing campaign.[39]

As a result of this article and criticisms from the families of fallen soldiers, the Australian government started to come under fire by the Australian people for unnecessary deaths of Australian troops and the now questionable Australian tactics.

Australians continued to push for additional involvement in the attack on Japan, but with the death of Curtin on July 5, 1945, the ties to MacArthur, who made many of the post-war decision in the Japanese occupation, weakened Blamey saw the need for Australian in volvement, and his political awareness caused him to create a shadow diplomatic service with agents in Washington and London, resulting in the ultimate recognition of the Australian role in the Southwest Pacific Theater.

The Potsdam Declaration declaring full and complete Japanese surrender came as a surprise to the Australian government because they were not invited to play a role in deciding the surrender or the terms of surrender after they had given so much to the fight in the Southwest Pacific Theater.[40] Australia then rose up and asserted itself. "Britain was informed that Blamey would represent the Australian government 'directly and not as attachment to your representative.'"[41] With the support of General MacArthur, General Blamey signed the Japanese declaration of surrender on September 2, 1945 as an Australian representative. An Australian, W. MacMahon Ball, was assigned to be the Commonwealth representative in the Allied Council for Japan, and the command of the Commonwealth Forces in the occupation of Japan was given to Australia, making it the first dominion government to command British forces, resulting in a movement toward sovereign equality of all Commonwealth members. Australia was recognized as a principal Pacific Power and an important part of the success of the Allied powers in the defeat of the Japanese by the British government.[42]

Bibliography

Primary

Johnston, George. "New Pacific Command." *The Mercury*, February 20, 1943. http://trove.nla.goau/newspaper/article/25946853?searchTerm=AmericansunderAustraliancommand&searchLimits=l-availability=yCCCl-australian=yCCCl-decade=194l-year=1943.

R.W.R. "Why Hold Blamey Responsible for Canberra's Hand?" *Pacific Islands Monthly*, April 17, 1945. https://nla.gov.au/nla.obj-315194580/view?sectionId=nla.obj-329343048&partId=nla.obj-315209331#page/n4/mode/1up.

Secondary

Australian War Memorial. *Field Marshal Thomas Albert Blamey*. https://www.awm.gov.au/collection/P10676218.

Australian War Memorial. *General Douglas MacArthur*. https://www.awm.gov.au/visit/exhibitions/underattack/end/macarthur.

Campbell, James. *The Ghost Mountain Boys: Their Epic March and the Terrifying Battle for New Guinea: The Forgotten War of the South Pacific*. Richmond, VA: James River Press, 2007.

Hagen, Jerome T. *War in the Pacific*, Vol. 3. Honolulu: Hawaii Pacific University, 2005.

Holzimmer, Kevin C. *General Walter Krueger, Unsung Hero of the Pacific War*. Lawrence, KS: University Press of Kansas, 2007.

Horner, D.M. *High Command: Australia and Allied Strategy 1939–1945*. Crows Nest, Australia: Allen & Unwin, 1982.

Johnston, George H. *The Toughest Fighting in the World, The Australian and American Campaign for New Guinea in World War II*. Yardley, PA: Westholme Publishing, 1943.

Manchester, William. *American Caesar, Douglas MacArthur 1880–1964*. Boston, MA: Little, Brown & Company, 1978.

Murray, Williamson and Allan R. Millett. *A War to be Won: Fighting the Second World War, 1937–1945*. Cambridge, MA: Harvard University Press, 2000.

National Archives of Australia. *Australia's Prime Ministers: John Curtin*. http://primeministers.naa.gov.au/primeministers/curtin/.

Prefer, Nathan. *MacArthur's New Guinea Campaign*. Madison, WI: Combined Books, 1995.

Robertson, John, and John McCarthy. *Australian War Strategy: 1939–1945*. St. Lucia: University of Queensland Press, 1985.

Shepard, Steven B. *American, British, Dutch, and Australian Coalition: Unsuccessful Band of Brothers*. Fort Leavenworth, KS: Command and General Staff College, 2003. http://www.dtic.mil/dtic/tr/fulltext/u2/a416342.pdf.

Spector, Ronald H. *Eagle Against the Sun: The American War with Japan*. New York: Vintage Books, 1985.

Taaffe, Stephen R. *MacArthur's Jungle War: The 1944 New Guinea Campaign.* Lawrence, KS: University Press of Kansas, 1998.

Westerfield, Hargis. *41st Infantry Division: The Fighting Jungleers.* Fredericksburg, VA: Billingsley Printing, 1980.

Notes

1 D.M. Horner, *High Command: Australia & Allied Strategy 1939–1945* (North Sydney, Australia: George Allen & Unwin Australia, 1982), 443.

2 Ibid., 443.

3 John Robertson and John McCarthy, *Australian War Strategy 1939–1945: A Documentary History* (St. Lucia: University of Queensland Press, 1985), 185.

4 Steven B. Shepard, *American, British, Dutch, and Australian Coalition: Unsuccessful Band of Brothers* (Fort Leavenworth, KS: Command and General Staff College, 2003), accessed May 18, 2017, http://www.dtic.mil/dtic/tr/fulltext/u2/a416342.pdf.

5 Horner, *High Command,* 187; William Manchester, *American Caesar: Douglas MacArthur 1880–1964* (Boston, MA: Little, Brown and Company, 1978), 214.

6 Ibid., 214.

7 National Archives of Australia, *Australia's Prime Ministers: John Curtin*, accessed October 20, 2017, http://primeministers.naa.gov.au/primeministers/curtin/.

8 Australian War Memorial, *General Douglas MacArthur,* accessed June 12, 2017, https://www.awm.gov.au/visit/exhibitions/underattack/end/macarthur.

9 Horner, 182.

10 Australian War Memorial, *Field Marshal Thomas Albert Blamey*, accessed October 20, 2017, https://www.awm.gov.au/collection/P10676218.

11 Nathan Prefer, *MacArthur's New Guinea Campaign: March-August 1944* (Madison, WI: Combined Books, 1995), 45.

12 Horner, 438.

13 Ibid., 439–40.

14 James Campbell, *The Ghost Mountain Boys: Their Epic March and the Terrifying Battle for New Guinea--The Forgotten War of the South Pacific* (New York: James River Press, 2007), 17.

15 Jerome T. Hagen, *War in the Pacific,* Vol. 3 (Hawaii Pacific University, HI, 2005), 105.

16 George H. Johnston, *The Toughest Fighting in the World, The Australian and American Campaign for New Guinea in World War II* (Westholme, Yardley, 1943), 171.

17 Stephen R. Taaffe, *MacArthur's Jungle War* (University Press of Kansas, KS, 1998), 54.

18 Campbell, *The Ghost Mountain Boys*, 224.

19 Nathan Prefer, *MacArthur's New Guinea Campaign, Taking the Outposts* (Combined Books, PA, 1995).

20 Horner, 291.

21 Hargis Westerfield, *41st Infantry Division: The Fighting Jungleers*, (Fredericksburg, VA: Billingsley Printing, 1980), 62.

22 Prefer, *MacArthur's New Guinea Campaign*, 19.

23 Kevin C. Holzimmer, *General Walter Krueger, Unsung Hero of the Pacific War* (Lawrence, KS: University Press of Kansas, 2007), 104.

24 George Johnston, "New Pacific Command," *The Mercury*, February 20, 1943. http://trove.nla.gov.au/newspaper/article/25946853?searchTerm=Americans%20under%20Australian%20command&searchLimits=l-availability=y%7C%7C%7Cl-australian=y%7C%7C%7Cl-decade=194%7C%7C%7Cl-year=1943.

25 Robertson and McCarthy, *Australian War Strategy*, 327.

26 Williamson Murray and Allan R. Millett, *A War to be Won: Fighting the Second World War*, (Cambridge: Harvard University Press, 2000), 199.

27 William Manchester, *American Caesar*, 298–99.

28 Ibid., 210.

29 Ibid., 427.

30 Ronald H. Spector, *Eagle Against the Sun: The American War with Japan* (New York: Random Books, 1985), 55–58.

31 Stephen R. Taaffe, *MacArthur's Jungle War: The 1944 New Guinea Campaign* (Lawrence, KS: University of Kansas Press, 1998), 17–18.

32 Ibid., 19–21, 27.

33 Gerhard L. Weinberg, *A World at Arms: A Global History of World War II* (New York: Cambridge University Press, 2005), 341.

34 Horner, 346–47.

35 Ibid., 374–75.

36 Taaffe, *MacArthur's Jungle War*, 54.

37 Borneman, *MacArthur at War*, 405–06.

38 Murray and Millett, *A War to be Won*, 492–93.

39 R.W.R., “Why Hold Blamey Responsible for Canberra’s Hand?” *Pacific Islands Monthly*, April 17, 1945, 5. https://nla.gov.au/nla.obj-315194580/view?sectionId=nla.obj-329343048&partId=nla.obj-315209331#page/n4/mode/1up.

40 Horner, 419–20, 430.

41 Ibid.

42 Ibid.

The Morality of Genocide: The Holocaust Revisited

Mike Rechtien
American Public University

Abstract

Subjective labels such as "evil" or "immoral" cannot effectively be evaluated, bringing little understanding to a phenomenon of human behavior called genocide. History clearly shows that the Holocaust was merely a single chapter in the ongoing saga of human prejudice-based mass destruction. The Bosnian genocide and the massacre of Jewish families by their fellow Polish townspeople in Jedwabne in 1941, for example, illustrate that this latent human impulse can be activated when three conditions are present: opportunity, impunity (perceived or actual), and moral basis.

Keywords: genocide, eugenics, holocaust, morality, Jedwabne, Bosnia, Polish, Nazi

La moralidad del genocidio: el Holocausto revisitado

Resumen

Las etiquetas subjetivas como "malvado" o "inmoral" no se pueden evaluar de manera efectiva, lo que genera poca comprensión de un fenómeno del comportamiento humano llamado genocidio. La historia muestra claramente que el Holocausto fue simplemente un capítulo en la saga en curso de destrucción masiva basada en prejuicios humanos. El genocidio bosnio y la masacre de familias judías por parte de sus conciudadanos polacos en Jedwabne en 1941, por ejemplo, ilustran que este impulso humano latente puede activarse cuando existen tres condiciones: oportunidad, impunidad (percibida o real) y base moral.

Palabras clave: genocidio, eugenesia, holocausto, moralidad, Jedwabne, Bosnia, polaco, nazi

种族屠杀的道德准则：重审大屠杀

摘要

例如“邪恶”或“不道德”等主观标签并不能被有效地评估，因此基本无助于理解被称为种族屠杀的人类行为现象。历史清晰显示，大屠杀仅仅是不断发展的、基于人类偏见的大型摧毁的长篇历史中的一个单一篇章。例如波斯尼亚种族屠杀和1941年耶德瓦布内市波兰人对犹太家庭展开的大屠杀都表明，这一潜在的人类冲动在三个条件存在的情况下能被激活：机会、免除惩罚（感知形式或实际形式）、道德基础。

关键词：种族灭绝，优生学，大屠杀，道德准则，耶德瓦布内，波斯尼亚，波兰人，纳粹

In the early 1990s, Dr. Donald Acheson led the World Health Organization's relief effort during the devastating wars in the former Yugoslavia. Reflecting on his time of service there, he recalled two critical conversations. The first took place with a Croat expatriate on his train ride across Europe and proved incomprehensible to him at the time. The gentleman made three points: 1) the war would end immediately with US military intervention; 2) without an effective peace initiative, the provision of humanitarian relief was immoral; and 3) allowing that the war would continue, he predicted that a series of massacres would occur in Bosnia. Acheson's second crucial conversation was after his arrival in Sarajevo with a retired schoolteacher who recounted his forty years of teaching in an integrated educational system where Croats, Jews, Muslims, and Serbs had equal rights and where he could recall no inter-communal violence. These two remarkably disparate statements caused Acheson to wonder how the first gentleman knew what few others understood at the time and was able to predict with such accuracy the genocidal tragedy about to overtake Bosnia, and what caused him to see Acheson's relief efforts as immoral.[1]

Half of a century after the Holocaust, and despite ongoing acts of extermination like the Bosnian genocide, it is generally held to be an inexplicable terror associated with a horrific era of German history. Often discussed and intensely researched, this genocide of European Jews is generally written off much too easily as an anomaly that simply could never happen again and ascribed to a madman and his temporarily crazed countrymen. Unfortunately, modern history clearly shows that it is merely a single chapter in the ongoing saga of human prejudice-based mass destruction. Since subjective labels such as "evil" or "immoral" cannot effectively be evaluated, they bring little understanding to this phenomenon of human behavior. And all the while, genocides continue; the Serbian genocide of the Bosnians, the Hutu genocide of the Tutsis, and the Sudanese genocide of local farmers in Darfur are just a few of the numerous atrocities committed in the more immediate past. The phenomenon proves itself to be a human issue, rather than one isolated to a time and culture from which we can comfortably distance ourselves.

Using evidence pertaining to the Holocaust, this paper argues that it was an example, albeit highly organized and efficient, of what is more likely a latent human impulse ready to germinate in certain specific climates; that the genocide of Jews during World War II was not only the responsibility of Nazi Germany or Germans in general; and that the traditional causative culprit of the Holocaust, the once highly popular science of eugenics and the policies it allegedly spawned, was less a cause than a product of this human impulse.[2]

Genocide

The term *genocide* was coined by Raphael Lemkin, a Polish-Jewish scholar fleeing Poland upon the Nazi invasion of 1939, as he reflected on the atrocities of the Ottoman Empire against Armenians, Greeks, and other Christians in the early twentieth century. In his 1944 book, *Axis Rule in Occupied Europe*, he explains it as race killing in all its various aspects. The aim is the annihilation of a group by destroying the essential foundations of their national culture. "Genocide," he writes, "is directed against the national group as an entity, and the actions involved are directed against individuals, not in their individual capacity, but as members of the national group." Four years later, in conjunction with the United Nations, the Genocide Convention gave greater focus to physical over cultural destruction in their draft, with an emphasis on "killing members of the [national, ethnic, racial, or religious] group." Over fifty years later, however, defining genocide stills remains a daunting task to many scholars, and preventing it proves nearly impossible. O'Lear and Egbert state that genocide is

> both persistent and evasive. It is persistent in that it seems to happen over and over, in different places, with different groups as perpetrators or victims but with horrifyingly similar trends in destructiveness and indifference to human suffering. It is evasive because one cannot discuss or think about genocide for very long without coming up against the perpetual problem of defining what, exactly, constitutes genocide.

For the purposes of this paper, genocide is discussed in the context of the known destructions of national, ethnic, and religious corporate entities (although even these designations are sometimes more perceptual than real, as can be argued in the case of the Hutus and Tutsis).[3]

Jedwabne, Poland, July 10, 1941

In 1736, after at least three hundred years of human settlement in the Polish hamlet of Jedwabne, Jews migrated to the area. By 1770, there were 387 Jews out of a population of 450, and by 1931, Jews accounted for a majority of the 2,167 inhabitants. In interwar Poland, Jews made up approximately 10 percent of the country's population, but composed nearly one-third of the Polish urban population, and the country was home to the second-largest overall number of Jews after the United States.[4]

Jedwabne was situated in the section of Poland annexed by the Russians following a secret agreement with Germany before the war. However, at the outbreak of war between Germany and Russia on June 22, 1941, the Russians pulled back to defensible lines and the Germans advanced into the territory. This day in June is given by some as the actual beginning date of the Holocaust, when concentration turned to extermination, and when the mass murders became standard practice. The Poles have long held that it was the Germans that committed these village-to-village slaughters as they moved into Polish territory vacated by Russia. But less than twenty days later, on July 10, 1941, the Polish citizens of Jedwabne viciously massacred their Jewish neighbors under the loose coordination of their mayor, Marion Karolak. This barbarity was a carry-over of similar events in the nearby towns of Wąsosz and Radziłów on preceding days. Witnesses testified that some of the Polish men who had massacred the Jewish men, women, and children in those hamlets came to Jedwabne that day to continue the pogrom. There was no attempt to stop the slaughter. It was carried out with impunity.[5]

In light of popular perception that the Holocaust was an inexplicable atrocity unique to the Germans of the early twentieth century, the Polish massacre of their fellow Jewish townspeople begs us to dig deeper. Since the Nazi Germans had just arrived in Poland in late June 1941, the Poles had not been subjected to years or even months of Third Reich ideology, propaganda, and enforcement. As Gross points out, the Poles "could not yet have managed to soak up the vicious anti-Jewish Nazi propaganda, even if they had been willing and ready." Additionally, the small number of Germans present in Jedwabne that day did not participate in the murder of the Jews, and even saved a few from slaughter to work in their outpost, limiting their actual participation to taking pictures. The Germans did give permission to the Poles, within an eight-hour time limit, to deal with the Jews as they wished. But their permission was not the same as the acts of torture, humiliation, beheading, clubbing, drowning, and burning of defenseless women, children, and men committed by the Polish townspeople. Nor was it an order to do so.[6]

The genocide lasted throughout the day in a small confined area of the township. It was impossible for anyone in the vicinity to have escaped not only the aftermath of the horrific event but also the atrocities as they were committed, according to Gross. Everyone, she states, "in possession of a sense of sight, smell, or hearing either participated in or witnessed the tormented deaths of the Jews of Jedwabne." And what they saw, smelled, and heard were their neighbors torturing and murdering other neighbors. It is imperative to attempt to understand the motivation for such actions which few humans would want to witness, much less carry out. It may well be true that the pogrom would not have occurred if the Germans had not been present to give permission and grant clemency; however, the actions of Polish citizens will not allow for the attribution of the genocide of the European Jews during the period known as World War II solely to the Nazis or strictly to the Germans. Since genocide did not begin or end with the Holocaust, this essay does not address the traditional questions of whether the Holocaust was ideologically driven by anti-Semitism and Nazi propaganda or whether it was a structural enterprise driven by human psychology and patterns of obedience to a recognizable authority that alleviates one from responsibility. Perhaps questions leading to such conclusions are too narrow in scope. The complexity of this issue can be evidenced by the inability of scholars to agree on a tangible explanation for such events, and the inability of world leaders to stop similar events, albeit smaller in scale and duration, from occurring. After all, is the slaughter of over 1,500 in eight hours in one small town in Poland any less terrible than of the 800,000 Tutsis in 100 days in Rwanda or the millions of Jews, gypsies, and infirmed over the course of several years in Nazi-occupied Europe? Each was driven by specific moral objectives and enacted when given opportunity with impunity.[7]

Morality

Morality can be defined as a system of beliefs concerning right and wrong. To label genocidal actions and their perpetrators as "immoral" does not allow for an understanding of the necessary moral basis for this ongoing human behavior. These events could not take place without a moral base. The failure to address this very issue is at the crux of the volatile state the world of humans often finds itself in. The perpetrators of genocide believe themselves to be morally right and in the service of protecting themselves, their families and their cultures from perceived injustices and threats—past, present, and future—and as we know from the study of psychology, perception is very much one's reality. Genocide is also an accepted religious principle for Jews and Christians alike, either carried out by God himself, as in the account of the destruction of Sodom and Gomorrah, or commanded by God of his followers to keep the chosen race morally pure. Through Moses, God commanded the Israelites to put to death not just people, but every living, breathing being in the cities that he commanded them to conquer.

> Only in the cities of these peoples that the LORD your God is giving you as an inheritance, you shall not leave alive anything that breathes. But you shall utterly destroy them, the Hittite and the Amorite, the Canaanite and

> the Perizzite, the Hivite and the Jebusite, as the LORD your God has commanded you, so that they may not teach you to do according to all their detestable things which they have done for their gods ... (Deut. 20:16-18; NAS)

Morality is a powerful learned perspective, often held to be absolute and unquestionable truth. Heroes and villains are defined from a moral perspective, but from an objective view, it is often a fine line divides them. Right and wrong are determined not by the action itself, but by a moral perspective. Morality thus dictates human social behavior; however, it is sometimes followed and sometimes not. In the case of the Israelites above, they disobeyed God by not fully carrying out his command and, as prophesied, lapsed into moral impurity—attributed to the influence of those who were spared.

Extermination of others by Christians, however, has been more common in our era. Take for example Pope Urban II's call in 1095 to wage war with the enemies of Christ in the Holy Land. In 1096, as prologue to the First Crusade, also known as the German Crusade, the Germans and French began their campaign with the massacre of Jewish communities in the Rhineland. This destructive social behavior could be viewed from a moral perspective as being sparked by the Pope's call to war, where a specific intention was generalized to all perceived enemies of Christ and, once again, when opportunity was joined with impunity. Just before the massacres, the Jews lived in peace and relative safety, while the moral perspective for genocide remained latent. There was no call to kill Jews, but there was an armed force faced with opportunity and perceived impunity. A latent, but powerful, shared morality was joined by opportunity and impunity through an authoritative call to action.[8]

What of the evidence for these factors on behalf of the Poles in Jedwabne and the surrounding hamlets? Opportunity and impunity have already been established. The German presence allows for that. But what of morality and the spark, or call to action, that ignited the massacres? If it were a known fact that the Germans had exterminated Jews in their section of occupied Poland, why did the Poles so eagerly volunteer for the job? There is no record of resistance (although it is well known that the Wyrzykowski family saved seven Jews). What moral principles were they enacting and from where did they originate? Was it from the same source as for the Germans, or was it the mere presence of the Germans themselves? And if the Germans had not been engaged in mass murder of Jewish men, women, and children in that area up to that point, as a proposed start date of June 22, 1941 suggests, what provided their spark for the implementation of the mass industrial extermination of Jews?

Eugenics

In the early twentieth century, eugenics was an internationally accepted field of scientific study. According to Field, it can be traced to a specific starting point with the published work of Francis Galton in 1865. Inspired by the 1859 publication of Darwin's *Origin of Species*, he associated the plasticity of the physical features of animals under supervision of a breeder with the inherited mental qualities of men. It coincided with the growing prominence of the masses and ignited a waning aristocratic regime from their "placid indifference." Eugenics grew into its heyday in the interwar period, building prominent followings in the scientific communities of Europe and the Americas, holding thirty-six international conferences with participants from fifty-six countries and officials from thirty nations. The movement claimed to have achieved sterilization laws in thirty-one of forty-eight states in the US. These laws became models for European eugenics policies. It appeared that this prominent new science was here to stay, as Fields attests in 1911:

> Unless many signs fail, the study of eugenics has established its claim to recognition among the hopeful applications of science in social reform ... and prompted investigations undertaken with a scholarly seriousness of purpose which bespeaks for them the critical estimate due to scientific work.

Eugenics' relationship to race hygiene and compulsive sterilization, and the perceived threat of overpopulation, drove policy formulation in many Western nations of the era. Interestingly, German eugenics laws and its policy of involuntary sterilization were described as a byproduct of earlier US and British eugenics activism and political marketing. Scandinavian countries also incorporated similar policies and had World War II not intervened, numerous countries, such as China, Russia, Poland, Australia, and South Africa, may also have followed suit. Therefore, a moral condemnation of German action in the Holocaust as a response to eugenics' history is sideways and dismissive. To understand eugenics as a theoretical justification for genocide, Gawin tells us, we must discuss the amount of responsibility for it that is attachable to an international phenomenon. If eugenics bears large responsibility, where is the evidence of the cessation of genocides that should have followed its demise into obscurity immediately after World War II?[9]

The relative peace in Europe in the late nineteenth century could be attributed to the Europeans' pillaging of resources and destruction of humans in their colonial territories. It gave them an opportunity for respite from war with each other. But in the early twentieth century, the world was increasingly viewed as a closed system due to colonialism and the expansion of Europeanism onto all available continents. The topic of overpopulation allowed long-held racist ideology to push nations toward control of reproduction and the need to focus on population *quality* instead of *quantity*. Most nations focused more on improvements of the individual through education and healthcare; in Germany, the primary focus was on the concept of race hygiene and the link between life and space known to Hitler and the German populace as *Leb-*

ensraum or "living space." This set the stage for Hitler's ideal war of conquest, subjugation, extermination, and expansion eastward, "where the German people would win for itself the [necessary] living space." This space was Poland, home to subhuman (*Untermenschen*) Slavs and the even biologically lower Jews. From there, Germany could expand its empire into Eastern European territories, preempt the threatening presence of 200 million Slavs, and exterminate and enslave the genetically flawed subhuman races to keep German "Aryan" blood pure. Perhaps this is well known to the reader, but what of the interwar Poles? They certainly would not have subscribed to this German belief concerning themselves. As eugenics was not a German phenomenon, how was it interpreted and enacted by the Poles? A better understanding of eugenics history may be helpful.[10]

Eugenics and demography were first blended in fascist Italy, where, in 1931, a fertility census was conducted. Fascism emerged in an interwar era of declining birth rates, political turmoil, and the Depression. Fascism is well known for its racist and genocidal ideologies, but these are not necessarily inherent in eugenics. Americans, the British, and Scandinavians saw strong lobbies and/or policy implementation of eugenic sterilization. Prophecies of the end of the white race and race suicide and fears of being overrun by colored races allowed fascist demographers to suggest more authoritarian solutions. Force was justified as an equivalent to natural process, whereas welfare was seen as opposed to natural selection. The latter was reserved for those deemed of high racial quality, and doctors and the social system were to serve the interests of race over the individual.[11]

This time period also corresponded with the condemnation of contraception in 1930 by the Catholic Church. Throughout that decade, many countries developed bans on abortion and/or subsidies for certain rural populations, such as the Gaelic speakers of Ireland. Italy and Spain developed their own ideology of a master race; however, both countries rejected policies of compulsory sterilization and extermination of ethnic minorities and the sick. In 1938, the Germans registered their total population and the census of 1939 allowed them to analyze the health and fertility of population cohorts. There was hope that hereditary studies and social controls could provide a solution to social issues like poverty or crime. Policies that other nations rejected were fully developed and implemented by the German regime, including compulsory abortion, sterilization, and castration and incentives for childbirth among pure blood Germans. These latter policies included stipends for avoidance of abortion or illegitimate childbirth and even having Adolph Hitler becoming godfather to a child in families with seven sons or nine total children. Hitler's population policies were the instruments of his racial ideals, especially when it came to the genocide of the Jews. *Mein Kampf* was published in 1925 and was widely distributed. In it, Hitler warned of racial degeneration through interbreeding as against the "iron laws" of nature. He also claimed the Aryan race as the only culture-bearing race and believed that it was the arrogance of the Jews that allowed them to deny racial law as natural law. He also addressed the concept of *Lebensraum* in *Mein Kampf*, but, again, the population pressure issue also shaped Italian, Japanese, and US policy and was not an invention of Hitler or fascism. This issue belonged to a worldwide collection of population experts using a spatial conception of the perceived problem. It grew out of intellectual input from historians and geographers of many nations. In both the Weimar Republic and Nazi periods, German geographers linked life and space and biography and geography and established a theory of geopolitics in which states were believed to be organic and unavoidably expansive. The state was held to be territory (space) and living (alive); thus the term *lebensraum* or "living space."[12]

Population policy and living space were on the mind and in the political discussions of many nations. Mussolini, in 1927, set a goal of 60 million Italians by 1950, and Spain's fascist Falange wanted 40 million Spaniards. Demographers offered endorsements of expansion for the Italians, Japanese, and Russians in places like Australia, New Zealand, Canada, and the Philippines. This ideology, along with the growing fears of race suicide, fit nicely into Hitler's comprehensive population policies. But what of the Poles? After nearly two years of Soviet occupation, with local elites being arrested or deported, private property being taken over, and vigorous secularization of religious institutions, the Poles welcomed the advancing Germans. The population was demoralized and brutalized. Add to the mix a license to use violence, as during the events in Jedwabne and surrounding areas, and the conditions were ripe for genocide. In carrying it out, the Poles could prove their value to the new regime, benefit from the material gain of the victims' possessions, and feel the satisfaction of settling a score for what they suffered. It proved enticing. Nearly half the adult men of Jedwabne were identified by name, supplied by fellow Polish citizens, as participants in the pogrom that day. Were they simply carrying out the desires harbored by most Poles?[13]

Polish historiography views the German extermination of Polish psychiatric patients as barbaric. Nonetheless, Gawin argues that eugenics supplied racism with scientific arguments and theoretical justification for the incitement of hatred for those perceived as weak and inferior, and that the Germans' *Final Solution* was not a product of traditional anti-Semitism as much as a combination of eugenics with anthropology and psychiatry. She delves into the state of Polish psychiatric practice in the interwar years, during the German occupation, and under the communist regime of the postwar era. These psychiatrists, as were all doctors, were conceptually committed to the health of their patients and also fully embraced the concept of *mental hygiene* toward their culture's most favorable mental health conditions. They were greatly influenced by their German counterparts' eugenics policy but preferred a softer approach, without anti-Semitic rhetoric. Almost to the man, they strongly supported the sterilization practices instituted in other nations, especially for alcoholics, drug addicts, and some of the mentally ill. Yet despite their best efforts and although the census at psychiatric hospitals continued to rise in the 1930s, those in the Polish government rejected forced sterilization outright, stating that doctors should advise ill persons not to marry but go no further.[14]

But with German occupation came the takeover of psychiatric hospitals by German medical personnel and the mass murder of the mentally ill by gun, poison, and starvation. It is highly likely that the Polish psychiatrists knew the Nazi executioners from international eugenics conferences. They were present during the extermination of their own patients, which came to light when the locals reported "patients groaning, praying, begging for mercy ... their terrible cries, and desperate attempts to escape before execution." And what became of these prominent Polish psychiatrists? Some committed suicide, some were shot along with their patients, and others were eventually killed by the Soviets. Some sixty perished in the war, and those that survived entered a half-century of shame and silence, never again revisiting their proud Polish eugenics tradition. The evidence points to the attempt, or at least the desire, to protect from death those they had proposed for sterilization. A strong interest in and knowledge of eugenics did not lead them to the same conclusions as their German counterparts. And although this discussion of Polish psychiatrists and those in their care differs in composition and content from the narrative of their fellow countrymen in Jedwabne, it does not differ in final outcome. The cries of agony and desperate attempts to survive in each narrative are strikingly familiar. But the perpetrators of the executions, Poles in one and Germans in the other, were not. In neither case could eugenic influence be appropriately assigned to the Poles involved. Other factors must be determined even as it pertains to the Holocaust. Eugenics was no more a cause of the destruction than was the club or the needle of the perpetrator.[15]

After the war, the once internationally popular eugenics movement, tied to fascism and genocide, went underground. No further international gatherings were held. One could say it was unique to Germany to carry these policies to the extent of the mass industrial genocide of the later Holocaust years, yet that assumes eugenics was the accelerant. Undoubtedly, Germany had the resources to carry out mass industrial extermination. It was not, however, only the Germans who were influenced by eugenics of the day—many other nations were as well, and yet did not commit genocide. Nor was it solely Germans who participated in the genocide of the European Jews, as is evidenced in the Jedwabne narrative. Eugenics may well have played an influencing role in an intricately complex foundation of racism, science, and opportunity, but it would be erroneous to label it as the cause. The common factors were the presence of the Nazis and the outcome of the victims—in the one case, mentally ill Poles and in the other, healthy and productive Jewish families. In both cases, the victims were well known to their fellow countrymen. If it were the morality of Nazi Germany that propelled their actions to horrific extremes, what of the Poles in Jedwabne when granted license and clemency with similar results?[16]

Genocide Revisited

Acheson, along with a growing number of scholars, believes that genocide is embedded in human social relationships and as such should be attended to within the scientific community. In each of the cases of genocide mentioned above, there was opportunity and impunity. To illustrate this point, there is documentation that Hitler laid out his plan for the genocide of the Poles and repopulation of their land by using, as supporting evidence, the silence present in the world community after the genocide of 1 million Armenians by the Turks during 1915–1916. If impunity provides opportunity, it should be considered to be a main ingredient of genocide and should guide the inquiry concerning not just the human atrocity of the Holocaust, but also the continual return to genocide inherent in modern human history. It is important to understand when just such an environment is being perceived. Secondly, an investigation into the constructs of individuals and cultures that prevent them from following suit when such an atmosphere is perceived should be undertaken. Gross attests to the complexity and the universality of the phenomenon:

> It is clear, from what happened in Jedwabne, that we must approach the Holocaust as a heterogeneous phenomenon. On the one hand, we have to be able to account for it as a system, which functioned according to a preconceived (though constantly evolving) plan. But, simultaneously, we must also be able to see it as a mosaic composed of discrete episodes, improvised by local decision-makers, and hinging on unforced behavior.

Concerning unforced behavior, a distinction should be made between a murderer in uniform as a state functionary acting under orders, possibly even with mental reservations about his actions, and a civilian killing their neighbor and his family of his own free will, as in the murderers in and around Jedwabne.[17]

Social Approach

Acheson believes genocide is the final stage in a three-tiered deterioration of a social relationship. He explains that there is a Group A, usually, but not always, the majority, which is in control, and a Group B, usually, but not always, the minority, which sees itself as the underdog. There are socioeconomic differences as well as differences of language, history, religion, ethnicity, and/or skin color. Group A views Group B as a challenge to its superior socioeconomic position and its cultural identity, which leads to insecurity. In turn, Group B is led to frustration and anger through its perception of exclusion from socioeconomic opportunity and from expressing its cultural heritage. This obvious first stage of discrimination, often including segregation, is nearly ubiquitous worldwide. Adding conscious and/or unconscious intimidation leads to the second stage – unplanned, sporadic, but often-cyclical violence. This is also quite common, but most often is the furthest extent of deterioration. The third and final step is the crucial factor – the open or secret involvement of the state or at least the sanction or perceived sanction of the controlling powers, as in the case of Jedwabne. This fuels public consent and gives control, however illusory, to the engaged participants of Group B to take retribution on Group A with impunity. The latent morality is ignited into action and given free rein to operate. And herein lies the issue of immoral behavior suggested by the unnamed gentleman on the train concerning the humanitarian relief effort in the former Yugoslavia. In Acheson's own words concerning his time in the affected area:

> Through all this we were enjoined to practice the strictest possible neutrality. This was essential in allowing us to pass freely across the front lines to support the health of refugees and other civilians whether they were Croat, Muslim, or Serb. Unfortunately, this neutrality was interpreted by some as imputing moral equivalence to aggressors and victims.

When the Republic of Bosnia and Herzegovina enlisted The International Court of Justice to hear their case against Serbia of the alleged genocide against Bosnian Muslims, the court found that the state of Serbia had not committed, conspired, incited, or been complicit in the genocide, but that it violated its obligation to prevent it. And concerning the humanitarian effort, Adam Jones, in his discussion of the *Challenges of Genocide Intervention*, states that in the aftermath the UN comprehended its part in enabling and even assisting genocide in Rwanda and Srebrenica, a most sobering realization.[18]

Genocide has been denounced as an international crime, but what to do about it remains difficult. The international community's lack of attention to its complexities and the general condemning of the perpetrators of the Holocaust as evil madmen serves to result in unpredicted interventionist outcomes that worsen the situation. Gross warns against such labeling:

> We must not assign collective responsibility. We must be clearheaded enough to remember that for each killing only a specific murderer or group of murderers is responsible.

In 2009, US scholar Alan Kuperman called Madeleine Albright and William Cohen's genocide prevention report "a recipe for failure," as it neglected, in his words, "the most profound lessons of past failures." His concerns were built on case studies where violent rebel groups enlisted propaganda and humanitarian appeals to provoke military intervention from the outside, often resulting in the encouragement of an otherwise doomed rebellion and fomenting further violence against civilians. The world community still has a lot to learn concerning the incredibly complex and evolving drama of human behavioral evolution. And as it is unlikely that any two genocidal episodes were preceded and unleashed by the same set of beliefs and circumstances, this topic deserves the most diligent attention of scholars, politicians, and world leaders.[19]

Conclusion

Eugenics cannot be blamed for genocide, much less the Holocaust. Terms such as "evil" and "immoral," applied to perpetrators like the Nazis, prove counterproductive in the attempt to understand the ongoing genocidal tendencies unleashed by groups of humans on other groups of humans from time immemorial to the present day. The fruitlessness of this approach can readily be seen by applying these labels to the Judeo-Christian god when instructing the ancient Jews to destroy every living creature in their new homeland. The reason given? Their immorality. The justification? Our morality. The likelihood that a nation's or religious body's god will be criticized or condemned by its faithful is close to nil. Furthermore, the Polish genocide of the Jews did not cease with the end of World War II. The most infamous postwar pogroms against Jews were in Kraków in 1945 and in Kielce in 1946 (although there is debate as to whether they were racially or politically motivated). What must be addressed is the underlying morality of national, ethnic, and religious groups. When joined with opportunity and impunity, a call to action or even just a backing off of political authorities can lead to genocide. And, as seen, relief efforts can add legitimacy to the carnage and fuel a dying flame.[20]

As epilogue, in 1996, Jewish author Aharon Appelfeld returned to the site of his former native village near Czernovitz where he lived until he was eight and a half. It brought up vivid memories of his childhood. Concerning June 1941, he is quoted in *Neighbors*:

> Who could imagine that in this village, on a Saturday, our Sabbath, sixty-two souls, most of them women and children, would fall prey to pitchforks and kitchen knives, and I, because I was in a back room, would manage to escape to the cornfields and hide?

Appelfeld hoped to find the grave of his mother. He inquired of the local villagers as to where the Jews were buried. He received no answer until someone he went to school with finally recognized him. Their silence soon proved to be concealment, not a lack of memory, as even the schoolchildren could point out the burial location. When he inquired as to its certainty, another villager assured him that it was the location—he was sixteen when he had helped bury the victims, his neighbors, there.[21]

Bibliography

Acheson, Donald. "Preventing genocide." *British Medical Journal* (December 7, 1996): 1415–17.

Amerasinghe, C.F. "The Bosnia Genocide Case." *Leiden Journal of International Law* (June 2008): 411–28.

Barrett, Deborah, and Charles Kurzman. "Globalizing Social Movement Theory: The Case of Eugenics." *Theory and Society* 33, no. 5 (October 2004): 487–527.

Bashford, Alison. "Nation, Empire, Globe: The Spaces of Population Debate in the Interwar Years," *Comparative Studies in Society and History* 49, no. 1 (January 2007): 170–201.

Bartov, Omer. *Germany's War and the Holocaust: Disputed Histories*. Ithaca: Cornell University Press, 2003.

Campbell, Bradley. "Genocide as Social Control." *Sociological Theory* 27, no. 2 (June 2009): 150–72. http://www.jstor.org/stable/40376129.

Field, James A. "The Progress of Eugenics." *The Quarterly Journal of Economics* 26, no. 1 (November 1911): 1–67. http://www.jstor.org/stable/1884524

Gawin, Magdalena. "Polish Psychiatrists and Eugenic Sterilization During the Interwar Period." *International Journal of Mental Health* 36, no. 1 (Spring 2007): 67–78. doi:10.2753/IMH0020-7411360106.

Gross, Jan T. *Neighbors: The Destruction of the Jewish Community in Jedwabne, Poland*. Princeton: Princeton University Press, 2012. Kindle.

Hankins, Frank H. "German Policies for Increasing Births." *American Journal of Sociology* 42, no. 5 (March 1937): 630–52.

Jones, Adam. "Challenges of Genocide Intervention." *Global Dialogue* (Winter 2013): 33–47.

Kuperman, Alan J. "Wishful Thinking Will Not Stop Genocide: Suggestions for a More Realistic Strategy." *Genocide Studies and Prevention* 4, no. 2 (August 2009): 191–99. http://scholarcommons.usf.edu/gsp/vol4/iss2/10.

Lemkin, Raphael. *Axis Rule in Occupied Europe: Laws of Occupation, Analysis of Government, Proposals for Redress*. Leiden: The Lawbook Exchange, 2008 [1944]. Chapter 9. http://www.preventgenocide.org/lemkin/AxisRule1944-1.htm.

McCullagh, C. Behan. "Bias in Historical Description, Interpretation, and Explanation." *History and Theory* 39 (February 2000): 39–66.

O'Lear, Shannon, and Egbert, Stephen L. "Introduction: Geographies of Genocide." *Space and Polity* 13, no. 1 (April 2009): 1–8.

Riley-Smith, Jonathon. *The Crusades, Christianity, and Islam*. New York: Columbia University Press, 2008.

Weindling, Paul, "Fascism and Population in Comparative European Perspective." *Population and Development Review 14*, Supplement: Population and Resources in Western Intellectual Traditions (1988): 102–21.

Notes

1 Donald Acheson, "Preventing Genocide," *British Medical Journal* (December 7, 1996): 1416.

2 Bradley Campbell, "Genocide as Social Control," *Sociological Theory* 27, no. 2 (June 2009): 151, http://www.jstor.org/stable/40376129; Acheson, "Preventing Genocide," 1415.

3 Adam Jones, "Challenges of Genocide Intervention," *Global Dialogue* (Winter 2013): 34–35; Raphael Lemkin, *Axis Rule in Occupied Europe: Laws of Occupation, Analysis of Government, Proposals for Redress* (Leiden: The Lawbook Exchange, 2008 [1944]): 79, Chapter 9, http://www.preventgenocide.org/lemkin/AxisRule1944-1.htm; Shannon O'Lear and Stephen L. Egbert, "Introduction: Geographies of Genocide," *Space and Polity* 13, no. 1 (April 2009): 1.

4 Jan T. Gross, *Neighbors: The Destruction of the Jewish Community in Jedwabne, Poland* (Princeton: Princeton University Press, 2012), Location 95, 310, Kindle.

5 Omer Bartov, *Germany's War and the Holocaust: Disputed Histories* (Ithaca: Cornell University Press, 2003), 4; Gross, *Neighbors*, location 471, 572, 637.

6 Gross, location 510, 660, 678, 1070.

7 Ibid., location 780, 1049; C. Behan McCullagh, "Bias in Historical Description, Interpretation, and Explanation," *History and Theory* 39 (February 2000): 49; Bartov, *Germany's War and the Holocaust,* 186.

8 Jonathon Riley-Smith, *The Crusades, Christianity, and Islam* (New York: Columbia University Press, 2008), 17.

9 James A. Field, "The Progress of Eugenics," *The Quarterly Journal of Economics* 26, no. 1 (November 1911): 1–4, http://www.jstor.org/stable/1884524; Deborah Barrett and Charles Kurzman, "Globalizing Social Movement Theory: The Case of Eugenics," *Theory and Society* 33, no. 5 (October 2004): 488, 509–10, 516; Magdalena Gawin, "Polish Psychiatrists and Eugenic Sterilization During the Interwar Period," *International Journal of Mental Health* 36, no. 1 (Spring 2007): 67. doi:10.2753/IMH0020-7411360106.

10 Alison Bashford, "Nation, Empire, Globe: The Spaces of Population Debate in the Interwar Years," *Comparative Studies in Society and History* 49, no. 1 (January 2007): 175, 191; Barrett and Kurzman, "Globalizing Social Movement Theory," 512, 517; Bartov,

4–5; Paul Weindling, "Fascism and Population in Comparative European Perspective," *Population and Development Review* 14, Supplement: Population and Resources in Western Intellectual Traditions (1988): 114.

11 Weindling, "Fascism and Population in Comparative European Perspective," 102, 109–10, 112, 197.

12 Ibid., 103, 106–07, 110; Frank H. Hankins, "German Policies for Increasing Births," *American Journal of Sociology* 42, no. 5 (March 1937): 632, 651; Bashford, "Nation, Empire, Globe," 175, 189.

13 Weindling, 114; Bashford, 189, 191; Gross, location 481, 764, 1378.

14 Gawin, "Polish Psychiatrists and Eugenic Sterilization," 67, 72, 75.

15 Ibid., 67–68, 76.

16 Barrett and Kurzman, 513; Gross, location 906.

17 Bradley, 151; Acheson, 1415; Gross, location 1075, 1137.

18 Acheson, 1415–17; C.F. Amerasinghe, "The Bosnia Genocide Case," *Leiden Journal of International Law* (June 2008): 411; Jones, 38.

19 Gross, location 1147; Alan J. Kuperman, "Wishful Thinking Will Not Stop Genocide: Suggestions for a More Realistic Strategy," *Genocide Studies and Prevention* 4, no. 2 (August 2009): 191, http://scholarcommons.usf.edu/gsp/vol4/iss2/10; Jones, 41.

20 Gross, location 1064.

21 Ibid., location 1064, 1083, 1087.

Revisiting the *Slaughter-House Cases* (1873)

Eric Balkan
American Public University

Abstract

Fear of yet another cholera epidemic plagued New Orleans in 1869. Attributing the cause of cholera to the pollution of the water supply, of which slaughterhouse waste dumping was a chief component, the Louisiana State Legislature passed a law regulating slaughterhouses. Opposition to this law by butchers eventually led to an 1873 Supreme Court ruling, which became a landmark decision in Fourteenth Amendment jurisprudence, as the butchers had sought relief under that amendment. The conventional modern opinion of Justice Samuel Freeman Miller's majority ruling in the *Slaughter-House Cases*, which denied the butchers' claim, is that it was an anti-Reconstruction ruling that gutted the "privileges or immunities" clause in the amendment, forcing future courts to rely on "substantive due process" to justify their decisions. However, relatively recently, several historians and legal scholars have offered a revisionist view that looks more favorably upon Miller's opinion, asserting that it was misinterpreted. This paper analyzes the majority and dissenting opinions in that case and related cases, reports on the congressional debate on the Fourteenth Amendment, and considers the historiography of the case, both conventional and revisionist. It concludes that Justice Miller, a Lincoln appointee and physician with a long-standing interest in public health, wrote a decision defending the actions of the biracial Louisiana Reconstruction legislature against white supremacists, defending it against laissez-faire economics, and defending the concept of federalism in general.

Keywords: slaughterhouse, Fourteenth Amendment, privileges or immunities, Reconstruction, cholera, Samuel F. Miller, butchers, Louisiana, revisionist, federalism, New Orleans

Revisando los *Casos de la Matanza* (1873)

Resumen

El temor a otra epidemia de cólera plagó la Nueva Orleans de 1869. Atribuyendo la causa del cólera a la contaminación del suministro de agua, del cual el vertido de desechos del matadero era un componente principal, la Legislatura del Estado de Luisiana aprobó una ley que regula los mataderos. La oposición a esta ley por parte de los carniceros finalmente llevó a un fallo de la Corte Suprema de 1873 que se convirtió en una decisión histórica en la jurisprudencia de la Decimocuarta Enmienda, ya que los carniceros habían buscado alivio bajo esa enmienda. La opinión moderna convencional del fallo mayoritario del juez Samuel Freeman Miller en los casos de Slaughter-House, que niega la afirmación de los carniceros, es que fue un fallo anti-Reconstrucción que destripó la cláusula de "privilegios o inmunidades" en la enmienda, obligando a los futuros tribunales a confiar en un "debido proceso sustantivo" para justificar sus decisiones. Sin embargo, relativamente recientemente, varios historiadores y estudiosos del derecho han ofrecido una visión revisionista que considera más favorablemente la opinión de Miller, afirmando que fue malinterpretada. Este documento analiza las opiniones mayoritarias y disidentes en ese caso, así como los casos relacionados, informa sobre el debate del Congreso sobre la Decimocuarta Enmienda, y considera la historiografía del caso, tanto convencional como revisionista. Concluye que el juez Miller, designado por Lincoln y médico con un largo interés en la salud pública, escribió una decisión defendiendo las acciones de la legislatura bi-racial de la Reconstrucción de Luisiana contra los supremacistas blancos, defendiéndola contra la economía de laissez-faire y defendiendo el concepto del federalismo en general.

Palabras clave: matadero, Decimocuarta Enmienda, privilegios o inmunidades, Reconstrucción, cólera, Samuel F. Miller, carniceros, Luisiana, revisionista, federalismo, Nueva Orleans

重审1873年的屠宰场案

摘要

1869年对另一场霍乱流行的恐惧笼罩着新奥尔良。将霍乱起因归咎于水供应污染（屠宰场废物倾倒是该污染的主要成分），路易安纳州立法机关通过了一项监管屠宰场的法案。

屠夫对该法案的反对最终导致1873年最高法院进行判决，该判决成为了第14条修正案法学中的一个标志性决定，因为此前屠夫企图在该法案下寻求帮助。对屠宰场案中法官塞缪尔·弗里曼·米勒（Samuel Freeman Miller）多数裁定法（否决屠夫所伸张的权利）所持的传统现代观点则是，这一裁定是一项将修正案中“特权或豁免权”条款加以提炼的反重建裁定，它强制要求未来法院依赖“实质性正当程序”来为各自的决定进行辩护。然而，相对而言近几年，一些历史学家和法律学者提出了一个修正主义者观点，这一观点看似更倾向于米勒的看法，并声称他的看法被错误解读。本文分析了屠宰场案例中的主流观点与反对观点，以及与国会就第14条修正案进行辩论的相关案例和报告，并考量该案例的历史学，包括传统与修正主义观点。本文结论认为，经林肯总统任命的法官米勒兼长期研究公共卫生的内科医生，写下了一个决定，维护路易安纳州重建时期双种族立法机关对反对白人种族优越论者所采取的行动，维护其对自由放任式经济学的反对，维护联邦主义的整体概念。

关键词：屠宰场，第14条修正案，特权或豁免权，重建，霍乱，塞缪尔·弗里曼·米勒，屠夫，路易安纳州，修正主义者，联邦主义，新奥尔良

In 1869, the Louisiana State Legislature passed a public health law that led to an 1873 Supreme Court ruling that became a landmark decision in Fourteenth Amendment jurisprudence. The conventional modern opinion of Justice Samuel Freeman Miller's majority ruling in the *Slaughter-House Cases* is that it was an anti-Reconstruction ruling that gutted the "privileges or immunities" clause in the amendment. However, relatively recently, historians and legal scholars have offered a revisionist view that looks more favorably upon Miller's opinion, asserting that it was misinterpreted. This paper shows that Justice Miller, a Lincoln appointee and physician with a long-standing interest in public health, wrote a decision defending the actions of the biracial Louisiana Reconstruction legislature against white supremacists, defending it against laissez-faire economics, and defending the concept of federalism in general.

New Orleans in 1869 was a city known for its squalid conditions.[1] Naturalist John James Audubon reportedly called the city's French Market the "dirtiest place in all the cities of the United States."[2] Probably the most noxious sites were the slaughterhouses.[3] Butchers had largely concentrated the slaughterhouses upstream from the city, in

Slaughterhouse Point.[4] They scattered others throughout the city, sometimes next to hospitals, schools, businesses, and tenements.[5] None of these locations were good, as the butchers simply threw the bloody waste from the slaughter of 300,000 animals a year into the streets or into the river, where it collected around the giant suction pipes from which New Orleans drew its water supply.[6] As far back as 1804, the government had attempted to move the slaughtering operations out of the city, but as the political power of the butchers grew, the slaughterhouses came back.[7] Epidemics of cholera and yellow fever occurred regularly, killing as many as forty thousand people in 1853 and thousands more into the 1860s.[8] These were part of global pandemics that still continue to kill thousands a year in developing countries with limited sanitation and clean water.

By 1869, other American cities had begun regulating slaughterhouses, including San Francisco, Boston, and Milwaukee.[9] New York City required the use of a centralized slaughterhouse, an idea unpopular with the city's butchers, but more popular with its residents.[10] The 1869 Louisiana law similarly required a central slaughterhouse.[11] The law banned the slaughtering of food animals within city limits, creating a new company with the exclusive (monopoly) privilege of operating a slaughterhouse.[12] A butcher could still conduct his trade, the law said, but must do his slaughtering at a specified place and pay a fee.[13]

Butchers in New Orleans attacked the law as a product of "black ignorance and Yankee (carpetbagger) greed."[14] As in other states, the legislature granted the monopoly privilege to wealthy individuals with political pull and to their friends in politics.[15] Critics complained of this corruption, charging those "carpetbaggers" and "scalawags" who benefited as being out for "plunder"—although historian Michael Les Benedict points out that the incorporators of the new Crescent City Company included both Republicans and Democrats.[16]

The law further annoyed white butchers by allowing freedmen to more easily become butchers, as they would no longer have to raise money to build their own slaughterhouses.[17] Primarily, though, the butchers, along with white New Orleans and the white press, were opposed to the biracial Reconstruction legislature on principle.[18]

The butchers unsuccessfully challenged the law in state court on the grounds that it was not within a state's police powers to charter a company for this purpose.[19] When the monopoly Crescent City Company got injunctions against the operation of a butchers' association slaughterhouse, the butchers appealed to federal court.[20]

The case—actually several similar cases lumped together as the *Slaughter-House Cases*—ended up before the Supreme Court. Lead attorney for the butchers was John A. Campbell (1811–1889), a former Associate Justice of the Supreme Court—where he had voted with the majority in *Dred Scott*—who left to become an Assistant Secretary of War for the Confederacy and was later a successful anti-Reconstruction attorney.[21] Campbell, on a mission to destroy Reconstruction state governments, feared the development of a democratic system incorporating blacks and uneducated immigrants.[22] He looked to the Fourteenth Amendment to keep this new population out of public life, by attempting to turn the amendment on its head.[23] Essentially, Campbell proposed to transfer all individual rights, especially economic rights, from the states to the federal government.[24] In one sense, Campbell would eventually get his wish: as the country abandoned Reconstruction, the state governments returned to a white rule that restricted the rights of blacks. This was a future, however, that neither side in the *Slaughter-House Cases* was able to foresee.

Campbell argued that the privileges and immunities of the Fourteenth Amendment—"No State shall make or enforce any law which shall abridge the privileges or immunities of citizens of the United States"—included all rights that people had, including common law rights preceding the Constitution, even going back to the Magna Carta.[25] Campbell also argued that the butchers were in de facto slavery to the monopoly Crescent City Company, violating the Thirteenth Amendment, and that their Fourteenth Amendment due process rights had been violated.[26]

The justice who would write the majority opinion, Samuel Freeman Miller (1816–1890), was from a much different background. He had lived in Keokuk, Iowa, a "Porkopolis," and had seen firsthand hog slaughtering and the town's attempts to regulate it.[27] Miller, a medical doctor, had written his medical school dissertation on the causes and treatment of cholera.[28] Historian Michael A. Ross calls Miller a pioneer in linking cholera to polluted water.[29]

The "privileges or immunities" issue that Campbell raised required Justice Miller to define just what privileges and immunities meant, as the amendment itself did not offer a definition. In the Congressional debates on the amendment, New Jersey Representative A.J. Rogers asserted that it included all the rights that citizens had come to enjoy, a sentiment similar to what Campbell would later argue in the case.[30] However, Constitutional scholar Akhil Reed Amar notes that Ohio Representative John Bingham, the key framer of the amendment, had specifically linked the privileges and immunities clause primarily to the Bill of Rights, in speeches and in a pamphlet he wrote.[31]

Michigan Senator Jacob Howard, sponsor of the amendment in the Senate, similarly linked them in a speech reprinted in *The New York Times* and in *The New York Herald*, the nation's best-selling newspaper in 1866.[32] During the debates, referring to the privileges and immunities clause in the fourth article of the Constitution, second section, Howard offered: "To these privileges, whatever they may be—for they are not and cannot be fully defined in their entire extent and precise nature—to these should be added the personal rights guarantied and secured by the first eight amendments of the Constitution."[33] Based on this, it appears that at least one of the framers had a fairly expansive view of these privileges and immunities, including what would later be called "incorporation"—that is, that the Bill of Rights, originally applying only to the federal government, would now act upon the states via the Fourteenth Amendment.

Yet, while the Fourteenth Amendment expanded on citizen rights, it was unclear to Congress just what that entailed. Bingham, for one, believed that this expansion did not infringe on existing states' rights, given that no state ever had the right to circumvent equal protection of the laws or privileges and immunities granted to all citizens.[34] Others thought that future Congresses would define them further. Indiana Senator Thomas A. Hendricks observed that he had never heard any senator actually define these rights.[35] Maryland Representative Charles E. Phelps stated his presumption that Congress, in the future, should determine what these immunities and privileges were to be.[36] Pennsylvania Representative Benjamin M. Boyer perhaps summed it up by saying that the amendment was "open to ambiguity and ... conflicting constructions."[37]

This ambiguity seems purposeful. Historian Eric Foner notes that Republicans resisted calls to define what "fundamental rights" of citizens, an oft-used phrase during the debates, actually were.[38] Foner goes on to say that the Fourteenth Amendment was intended as a broad statement of principle that disavowed any "legal discrimination" and enlarged the concept of freedom for all citizens.[39] But constitutional scholar David S. Bogen writes that what the amendment's framers meant to say is "slippery," with too many possible interpretations.[40] Kevin Christopher Newsom, a former appellate litigator and current United States Appeals Court judge, sees limits to privileges and immunities, noting that Bingham, in an 1871 Congressional speech, saw these rights as being chiefly limited to the first eight amendments in the Bill of Rights.[41]

Congress seemed to agree that the purpose of the amendment was to ensure the equality of rights among all citizens. Vermont Senator Luke P. Poland noted that the first clause did not go beyond the original provision in Article IV of the Constitution, that "the citizens of each State shall be entitled to all privileges and immunities of citizens in the several States."[42] New York Representative Henry J. Raymond noted that the amendment followed his long-time support of providing equal rights to all.[43] Representative Phelps added: "[the amendment] allows Congress to correct the unjust legislation of the States, so far that the law which operates on one man shall operate *equally* upon all."[44] (The reference to unjust legislation was to the purposefully discriminatory Black Codes.)

The Court decided the case five to four against Campbell and the butchers. Justice Miller wrote the majority opinion, joined by Justices Nathan Clifford, David Davis, William Strong, and Ward Hunt. Three of the four dissenters wrote their own separate opinions: Stephen J. Field, Joseph P. Bradley, and Noah H. Swayne; Chief Justice Salmon P. Chase simply joined the other dissenters. This split was not along party lines, as, for

Death's Dispensary by British cartoonist George Pinwell, referencing the recent discovery by Dr. John Snow that cholera was spread through contaminated drinking water. Source: University of Florida Digital Collections, https://ufdc.ufl.edu/UF00078627/00010/206j.

Author of the majority opinion in the Slaughter-House Cases, Supreme Court Associate Justice Samuel Freeman Miller (1816–1890, photographed by Mathew Brady. Source: The United States Library of Congress Prints and Photographs Division, digital ID cwpbh.03994.

Lead attorney for the plaintiff butchers, former Supreme Court Justice John Archibald Campbell (1811–1889), photographed by Matthew Brady. Source: The United States Library of Congress Prints and Photographs division, digital ID cwpbh.04017.

example, Miller and Chase were both Lincoln appointees. But Lincoln had picked justices based on unionism and not on economic philosophy.[45] Miller, for instance, a product of a small-town Western upbringing opposed to the financial and railroad empires headquartered back east, found himself left out in a Republican Party increasingly tied to big business.[46]

Miller rejected each of Campbell's arguments. The existence of a monopoly, no matter if it were a good idea, Miller ruled, did not prevent the butchers from exercising their trade.[47] Furthermore, he found that the police power of the states allowed regulation.[48] Miller also ruled against the butchers' Fourteen Amendment claim that this monopoly had effectively taken their property without due process of law.[49]

The heart of Miller's opinion, and the part that became the most controversial, revolved around the "privileges or immunities" clause. Miller approached this by first finding that Americans have dual[50] citizenship, and so dual rights, in the United States and in the state they reside in. The Fourteenth Amendment, Miller found, dealt only with the rights of a United States citizen.[51] Other rights, under the concept of federalism, belonged to the states to administer.[52] So, it is the states that provide the right to exercise a trade, and not the federal government. To read the Fourteenth Amendment otherwise, to include under federal domain all of a citizen's rights, including "common rights" that existed before the Constitution—which is what the plaintiffs were asking for—would radically disrupt the whole concept of federalism that governed the relationship of the state and federal governments to each other.[53] Miller did not believe the framers of the post-war amendments, or the state legislatures that ratified them, had intended this.[54] Miller firmly believed that the rationale for these amendments was "the freedom of the slave race."[55] Miller mentioned nothing one way or the other about incorporation—the concept that the Bill of Rights applies to the states—because no right enumerated in the Bill of Rights was at issue in this case. Finding that the rights being argued in the case were rights of state citizens, to be decided by state governments, he found no need to define the "privileges and immunities" of United States citizens.[56] Conventional historiography criticizes Miller's decision not to list those rights, but it is legally understandable, as it was not his place in writing this particular case opinion to decide what they were.

Having read the Fourteenth Amendment as continuing the concept of dual citizenship, Miller then read the next sentence: "No State shall make or enforce any law which shall abridge the privileges or immunities of citizens of the United States" as applying only to those rights attached to national citizenship and not to state citizenship.[57] Miller feared that, without federalism, as a practical matter, the federal courts would be "inundated with legislation."[58]

Four members of the Court found Campbell's arguments to be reasonable and so dissented from the majority opinion. This began a movement towards a pro-business (laissez-faire) interpretation of the Fourteenth Amendment, using the concept of "substantive due process," including the classic right-of-contract case, *Lochner v. New York* (1905).[59] (Substantive due process is a legal doctrine intended to prevent the violation of rights even when proper due process procedures have been followed.) In particular, Justice Field's dissenting opinion was the polar opposite of Miller's. He wrote that the amendment was intended to protect "common rights" from the States.[60] He further noted that if the monopoly were sustained, then there would be no limit to favors by a state government to preferred businesses.[61] Field also rejected the concept of United States citizenship apart from state citizenship.[62] "The privileges and immunities designated are those which of right belong to the citizens of all free governments," taking a lead from Justice Washington in *Corfield v. Coryell* (1823).[63] Field added: "the right of free labor" is "one of the most sacred and imprescriptible rights."[64] Essentially, Field argued that the amendment had been adopted in order to set the federal government as the protector of Americans' common rights."[65] From another point of view, Field may have viewed the slaughterhouse law as an attempt at redistribution—agreeing with the butchers' complaint that their property had been taken from them—rather than as a health issue.[66]

Justice Bradley, while generally agreeing with Justice Field, added some further thoughts.[67] He, too, believed that there are fundamental rights belonging to the citizens of any free government.[68] To Bradley, state citizenship was secondary.[69] Citizens had a right to choose any lawful employment.[70] He found the Louisiana law to be "arbitrary and unjust" and not a proper exercise of the police powers. The granting of monopolies "is an invasion of the right of others to choose a lawful calling, and an infringement of personal liberty."[71] In answer to Miller's fear of the collapse of federalism, Bradley wrote that under his interpretation, he did not believe the federal government would end up supervising the internal affairs of the states.[72]

Justice Swayne, in his own dissent, added that monopoly abridges both the privileges and immunities of a citizen and deprives him of liberty or property without due process of law.[73] Unlike Miller, Swayne was not afraid of a radical change in government.[74] To Swayne, every state citizen was a United States citizen and all have equality, protected by the Fourteenth Amendment, in state-created privileges and immunities.[75]

Radical Republicans and western Democrats criticized Miller's decision as too limited in its interpretation of the Fourteenth Amendment.[76] Legal scholar Charles Warren notes that Vermont Senator George F. Edmunds, who had helped pass the amendment, thought Miller's decision a "radical" departure from what the amendment's framers had intended or what the amendment actually said.[77] Warren also notes that New York Senator Roscoe Conkling, leader of the Stalwart anti-reform branch of the Republican Party and a member of the House at the time it passed the amendment, complained that the amendment was not only meant to help the freed slaves but also protect business from legislators.[78]

Constitutional law scholar Richard L. Aynes notes contemporary criticism of the majority opinion from the *Chicago Legal News* in 1873 and Sedgwick's *Statutory and Constitutional Law* in 1874, on grounds that Miller was reading the Fourteenth Amendment as being passed primarily to help "the colored citizen," while they believed it was intended generally.[79] Other publications were less subtle. *The Cincinnati Enquirer*, on April 16, 1873, called out the "degeneracy of the Court" in upholding "a law passed by a so-called Legislature, elected by the bayonet and through the agency of the most degraded and ignorant portion of the population ... to reward particular favorites."[80] In a similar vein, the *Southern Law Review* claimed that the decision sustained a "menacing monopoly created by a corrupt and ignorant carpet-bag State Government."[81]

Still, Warren notes that many Americans, particularly those leery of government centralization, did not share this attitude.[82] *The Chicago Tribune*, *The Nation*, and *The Independent*, among others, defended the decision.[83] *The New York World*, April 16, 1873, found Miller had "fairly interpreted" the new amendments intention to retain the concept of federalism.[84] *The New York World* went on to say that "certain shallow people" have "gone crazy" about the scope of the Fourteenth Amendment.[85]

At least two later cases that cited *Slaughter-House* reiterated Miller's defense of federalism. In *In re Kemmler* (1890), Chief Justice Melville Fuller wrote for the majority that the privileges and immunities of citizens of the United States, granted by the Constitution, are protected by the Fourteenth Amendment, but that the "protection to life, liberty, and property rests, primarily, with the states."[86] In *Twining v. New Jersey* (1908), Justice William Henry Moody wrote for the majority that while the decision in *Slaughter-House* had disappointed those looking for the Fourteenth Amendment to have greater effect, the alternative would have been much worse.[87] If the dissenters had prevailed, then the authority of the states would have been much diminished, by "subjecting all their legislative and judicial acts to correction by the legislative and review by the judicial branch of the national government."[88] He added that the judgment in *Slaughter-House* continues to be upheld.[89]

Since then, historians and legal scholars have come down hard on Miller's decision, especially its lack of clarity regarding incorporation and its limits on privileges and immunities. Amar writes that the conventional reading of Miller's opinion "falls far short of incorporation."[90] He further refers to it as "strangling the privileges or immunities clause in its crib."[91] Lawrence Tribe writes that the *Slaughter-House Cases* "artificially constricted" the Fourteenth Amendment's "privileges or immunities" clause by reading those words more narrowly than how most Americans understood them or how the amendment's framers intended them, which was to protect citizens' rights against the states.[92] Benedict sums up Miller's opinion: Miller "concluded that the 'one pervading purpose' behind the Civil War amendments was to secure the freedom of black Americans, not to expand or add protections for the rights of white."[93] (However, Miller also wrote: "We do not say that no one else but the negro can share in this protection."[94]) Legal historian Donald G. Nieman adds: "The Slaughterhouse Cases dealt the privileges and immunities clause a blow from which it never recovered."[95]

Foner also expresses disappointment with the limits of the decision. He views Miller's definition of rights due to "national citizenship" as too narrow to be of any use to blacks.[96] Essentially, Foner is looking for the federal government to protect the rights of blacks against private individuals and/or local officials not enforcing the law,[97] and not just against racist state laws; however, the Supreme Court rejected this concept in the 1883 *Civil Rights Cases*, with both Miller and Bradley on the same side that time.[98]

Ross finds that the current consensus on Miller's interpretation of the Fourteenth Amendment's privileges and immunities clause is that it was incorrect, ignored the intent of the framers, and did not incorporate the Bill of Rights.[99] The criticism about incorporation is particularly critical, as it is the principle of incorporation that has largely defined late twentieth century constitutional law, from Miranda rights to abortion rights. Ross notes the overall dissatisfaction with the opinion, quoting a Supreme Court historian, Leo Pfeffer, that "the only thing slaughtered in the Slaughterhouse cases was the right of the Negro to equality."[100] Ross concludes that the majority opinion and Miller have become "lightning rods for censure in the tragic unraveling of Radical Reconstruction."[101]

Aynes, for one, reads Miller's opinion as being antagonistic to incorporation, while seeing Congressional views in the debate over the 1875 Civil Rights Act as supporting incorporation.[102] He thinks the amendment was being "frittered away by judicial construction," quoting former Attorney General Benjamin Bristow (1873), while former Confederate and future Supreme Court Justice Lucius Q.C. Lamar saw the Court, especially regarding *Slaughter-House*, as anti-Reconstruction.[103]

Aynes further notes that historian John W Burgess and others in the 1890s saw *Slaughter-House* as non-incorporationist.[104] He explains Miller's decision with a reference to legal scholar Christopher G. Tiedman, who, in 1890, thinking that a literal reading of the Fourteenth Amendment would have ended local self-government in the US, applauded the Miller opinion, even though it violated the intent of both the framers and the ratifiers.[105] Tiedman believed, according to Aynes, that Miller's opinion opposed the popular will at the time, but was more in line with what cooler heads would want.[106] Aynes also finds that Congressional opinion had changed by 1873, so that Miller's opinion was in line with that thinking, just not in line with the Congress that had passed the Fourteenth Amendment in 1866.[107] This is similar to what Warren said, back in 1923, that by 1873, the country had moved away from centralization and back to the states' rights views held by Democrats before the war.[108]

In recent years, scholars have taken a fresh look at the *Slaughter-House Cases* and especially its effect on incorporationism. In an in-depth article on the topic, Newsom finds that the conventional reading of the case is not unreasonable, but is also not the best.[109] Like conventional historiographers, Newsom sees the intent of the amendment framers as incorporating the Bill of Rights, but does not think that Miller interfered with this.[110] He finds nothing in Miller's opinion to think he was opposed to incorporation.[111] Miller had specifically said that federal privileges and immunities include the "right to peaceably assemble and petition for redress of grievances."[112] This is from the First Amendment, so it does look like he intended to incorporate at least some of the Bill of Rights.[113] While conventional historiography pointed to the skimpy list of privileges and immunities that Miller had listed, Newsom views this list as illustrative rather than exhaustive.[114]

Newsom finds Miller thinking that Congress had designed the Fourteenth Amendment to promote equality in civil (economic) rights and not to expand them.[115] Newsom thinks that Justice Field misread the Civil Rights Act of 1866, thinking it expanded rights, whereas Newsom thinks it just ensured that blacks had the same rights as whites, whatever those rights were.[116]

Ross, despite the conventional view of Miller's opinion, saw it as upholding the authority of a biracial Reconstruction government to deal with a health crisis in New Orleans, and in so doing prevent conservatives like Justice Field from putting private property outside of state regulation.[117] Ross finds that Miller and the majority did not believe that Congress had designed the Fourteenth Amendment to do away with the traditional system of federalism,[118] which would have opened the gates to federal overruling of state business regulations.[119] Ross believes that Miller was afraid of the "notoriously conservative" federal judiciary striking down state legislation.[120] Further, it was not clear at this time that Reconstruction would fail, and so Miller had faith in state governments as protecting the rights of blacks.[121] Miller felt, according to Ross, that Congress passed the Fourteenth Amendment as an antidote to the Black Codes, but not something that would strike down a state health regulation having nothing to do with race.[122]

Keeping the case in historical perspective is critical to understanding it, including what Miller's contemporaries thought about incorporation. Professor Bryan H. Wildenthal, after studying the evidence, believes that the *Slaughter-House* majority adopted a reading of the Fourteenth Amendment that applied, at a minimum, the Bill of Rights to the states—and that this view was widely shared at the time.[123] He finds that disincorporation happened afterwards, in cases like *Plessy v. Ferguson* (1896), the landmark racial segregation case.[124]

One of the strikes against Miller is that there were later cases where the Court did not invoke incorporation, but, as Newsom points out, none of these cases actually involve the incorporation issue.[125] This is an important point, often overlooked, because, typically, an appellate court will only consider legal issues raised by the parties, and not address issues that the parties could have mentioned, but did not. For instance, *US v. Cruikshank* (1874) was about private actions of individuals and not state actions.[126] In *Eilenbecker v. District Court* (1890) and *Hurtado v. California* (1884), appellants made the legal error of invoking the Bill of Rights against state actions without invoking the Fourteenth Amendment, so the issue of incorporation never came up.[127]

The Court eventually followed the dissenters, except that instead of using the privileges and immunities clause, it used "substantive due process," overruling state regulation of business in *Allgeyer v. Louisiana* (1897) and, in the case that gave the court of this period its name, *Lochner*.[128] But this "ultimate victory of virulent racism and laissez-faire jurisprudence is not what Miller and the Court's majority in the *Slaughter-House Cases* intended."[129]

Modern critics have often castigated Justice Miller's opinion in *Slaughter-House*, thinking that it was responsible for future courts not being able to use the Fourteenth Amendment's "privileges and immunities" clause to prevent rights violations. However, that view could well be unfair. *Slaughter-Houses Cases* depict a state, through its Reconstruction legislature, trying to provide equal protection for blacks. That this situation would later reverse, with states preventing equal rights while the federal government moved to protect them, is not something Miller could have foreseen. Miller was also handicapped by the framers of the Fourteenth Amendment, trying to negate states' Black Codes, not considering, back in 1868, that the big problem for blacks would be private action rather than direct state action. However, Miller's decision did not prevent Congress or the Courts from enforcing equal rights, including the use of incorporation. That they did not do so cannot be laid on *Slaughter-House*.

The majority's commitment to federalism created a defense of an action of a Reconstruction state government against both white supremacist opposition and oversight by a federal government favoring laissez-faire economics. Miller accomplished this by differentiating between the rights of citizens of a state and the rights of a citizen of the United States. With that distinction, the "privileges or immunities" of the Fourteenth Amendment only applied to rights given to citizens of the United States by the Constitution or Congress, and not to common rights that traditionally belonged to the states to uphold. Later courts, the laissez-faire "Lochner" courts, got around this by defending common rights, like the right of contract, using the due process clause in the amendment. Later still, during the Great Depression, Congress and the Courts would reject that interpretation. Eventually, they would act against racism by private interests, primarily using the Commerce Clause (US Constitution, Art. 1, Sec. 8, cl. 3.), but that would only come when much of the public had begun to see racism as unacceptable.

Bibliography

Amar, Akhil Reed. *America's Constitution: A Biography*. New York: Random House, 2005.

———. "The Bill of Rights and the Fourteenth Amendment," *The Yale Law Journal* 101 (April 1992): 1193–284. http://www.jstor.org/stable/796923.

Aynes, Richard L. "Constricting the Law of Freedom: Justice Miller, the Fourteenth Amendment, and the Slaughter-House Cases." *Chicago-Kent Law Review* 70 (1994): 627–95. https://scholarship.kentlaw.iit.edu/cklawreview/vol70/iss2/8.

Benedict, Michael Les. "Slaughter House Cases." In *Oxford Companion to the Supreme Court of the United States*. 2nd ed. Edited by Kermit L. Hall, 922–24. New York: Oxford University Press, 2005.

Bogen, David S. "Slaughter-House Five: Views of the Case." *Hastings Law Journal* 55 (December 2003): 333–85. https://repository.uchastings.edu/cgi/viewcontent.cgi?article=3544&context=hastings_law_journal.

"Congressional Debates of the 14th Amendment." *Congressional Globe*, 39th Congress (1866) First Session, 1–4312. https://www.scribd.com/document/36527058/Congressional-Debates-of-the-14th-Amendment?doc_id=36527058&download=true&order=438384759.

Foner, Eric. *Reconstruction: America's Unfinished Revolution, 1863–1877*. Rev. ed. New York: HarperCollins Publishers, 2014.

Kens, Paul. *Lochner v. New York: Economic Regulation on Trial*. Lawrence, KS: University Press of Kansas, 1998.

Nieman, Donald G. "The Fourteenth Amendment Receives Its First Judicial Construction". In *Historic U.S. Court Cases: An Encyclopedia*. 2nd ed. Edited by John W. Johnson, 469–77. New York: Routledge, 2001.

Newsom, Kevin Christopher. "Setting Incorporationism Straight: A Reinterpretation of the Slaughter-House Cases." *The Yale Law Journal* 109, no. 4 (Jan 2000): 645–744.

Ross, Michael A. "Justice Miller's Reconstruction: The Slaughter-House Cases, Health Codes, and Civil Rights in New Orleans, 1861–1873." *The Journal of Southern History*, 64, no 4 (Nov 1998): 649–76. http://www.jstor.org/stable/2587513.

———. *Justice of Shattered Dreams: Samuel Freeman Miller and the Supreme Court during the Civil War Era*. Baton Rouge: LSU Press, 2003.

Tribe, Lawrence. *The Invisible Consti-*

tution. New York: Oxford University Press, 2008.

Warren, Charles. *The Supreme Court in United States History*. Vol. 3. 1923. Reprint, New York: Cosimo Classics, 2011.

Wildenthal, Bryan H. "How I Learned to Stop Worrying and Love the *Slaughter-House Cases*: An Essay in Constitutional-Historical Revisionism." *Thomas Jefferson Law Review* 23, no. 2 (Spring 2001): 241–48.

Notes

1 Michael A. Ross, "Justice Miller's Reconstruction: The Slaughter-House Cases, Health Codes, and Civil Rights in New Orleans, 1861-1873," *The Journal of Southern History* 64, no. 4 (November 1998): 653, http://www.jstor.org/stable/2587513.

2 Ibid.

3 Ibid., 654.

4 Ibid.

5 Ibid.

6 Ibid.

7 Ibid., 654–55.

8 Ibid.

9 Ibid., 655.

10 Ibid.

11 Ibid., 656.

12 *Slaughter-House Cases*, 83 US 36, 59–60 (1873).

13 Ibid., 61.

14 Michael A. Ross, *Justice of Shattered Dreams: Samuel Freeman Miller and the Supreme Court during the Civil War Era* (Baton Rouge: LSU Press, 2003), 200.

15 Michael Les Benedict, "Slaughter House Cases," in *Oxford Companion to the Supreme Court of the United States,* 2nd ed., ed. Kermit L. Hall (New York: Oxford University Press, 2005), 922.

16 Ibid.

17 Ross, "Justice Miller's Reconstruction," 656.

18 Ibid., 661.

19 Benedict, "Slaughter House Cases," 923.

20 Ibid.

21 Ross, "Justice Miller's Reconstruction," 665.

22 Ibid., 199.

23 Ibid.

24 Kevin Christopher Newsom, "Setting Incorporationism Straight: A Reinterpretation of the Slaughter-House Cases," *The Yale Law Journal* 109, no. 4 (Jan 2000): 661.

25 Richard L. Aynes, "Constricting the Law of Freedom: Justice Miller, the Fourteenth Amendment, and the Slaughter-House Cases," *Chicago-Kent Law Review* 70 (1994): 634, https://scholarship.kentlaw.iit.edu/cklawreview/vol70/iss2/8.

26 Ibid.

27 Ross, *Justice of Shattered Dreams*, 202.

28 Ross, "Justice Miller's Reconstruction," 668.

29 Ibid.

30 "Congressional Debates of the 14th Amendment," *Congressional Globe*, 39th Congress (1866) First Session, 361, https://www.scribd.com/document/36527058/Congressional-Debates-of-the-14th-Amendment?doc_id=36527058&download=true&order=438384759.

31 Akhil Reed Amar, *America's Constitution: A Biography* (New York: Random House, 2005), 387.

32 Ibid., 387–88.

33 *Congressional Debates*, 12.

34 Ibid., 371.

35 Ibid., 162.

36 Ibid., 232.

37 Ibid., 294.

38 Eric Foner, *Reconstruction: America's Unfinished Revolution, 1863–1877*. Rev. ed. (New York: HarperCollins, 2014), 257.

39 Ibid., 257–58.

40 David S. Bogen, "Slaughter-House Five: Views of the Case," *Hastings Law Journal* 55,

no. 2 (December 2003): 384, https://repository.uchastings.edu/cgi/viewcontent.cgi?article=3544&context=hastings_law_journal.

41 Newsom, "Setting Incorporationism Straight," 698.

42 Congressional Debates, 101.

43 Ibid., 317.

44 Ibid., 276.

45 Ross, *Justice of Shattered Dreams*, xviii.

46 Ibid., xix.

47 *Slaughter-House Cases*, 60.

48 Ibid., 62.

49 Ibid., 61.

50 Ibid., 74.

51 Ibid.

52 Ibid.

53 Ibid., 78.

54 Ibid.

55 Ibid., 71.

56 Ibid., 78-79.

57 Newsom, 653.

58 Ibid., 708.

59 Paul Kens, *Lochner v. New York: Economic Regulation on Trial* (Lawrence, KS: University Press of Kansas, 1998), 103.

60 *Slaughter-House Cases*, 89.

61 Ibid., 93.

62 Ibid., 95.

63 Ibid., 97.

64 Ibid., 110.

65 Ibid., 93.

66 Ross, *Justice of Shattered Dreams*, 205.

67 *Slaughter-House Cases*, 111.

68 Ibid., 114.

69 Ibid., 112.

70 Ibid., 113.

71 Ibid., 120.

72 Ibid., 123.

73 Ibid., 128.

74 Aynes, "Constricting the Law of Freedom," 643.

75 Ibid.

76 Charles Warren, *The Supreme Court in United States History*, Vol. 3 (1923, Reprint New York: Cosimo Classics, 2011), 263.

77 Ibid.

78 Ibid.

79 Aynes, 645.

80 *Cincinnati Enquirer*, April 16, 1873.

81 Warren, *The Supreme Court in United States History,* 264.

82 Ibid., 264–65.

83 Ibid., 264.

84 Ibid., 265.

85 Ibid.

86 *In re Kemmler,* 136 US 436, 448–49.

87 *Twining v. New Jersey*, 211 US 79, 96–97.

88 Ibid.

89 Ibid.

90 Akhil Reed Amar, "The Bill of Rights and the Fourteenth Amendment," *The Yale Law Journal* 101, no. 6 (April 1992): 1259, http://www.jstor.org/stable/796923.

91 Ibid.

92 Lawrence Tribe, *The Invisible Constitution* (New York: Oxford University Press, 2008), 44.

93 Benedict, 924.

94 *Slaughter-House Cases*, 72.

95 Donald G. Nieman, "The Fourteenth Amendment Receives Its First Judicial Construction," *Historic U.S. Court Cases: An Encyclopedia,* 2nd ed., ed. John W. Johnson (New York: Routledge, 2001), 477.

96 Foner, *Reconstruction,* 531.

97 Ibid.

98 *Civil Rights Cases*, 109 US 3, 24.

99 Ross, "Justice Miller's Reconstruction," 650.

100 Ibid.

101 Ibid., 649.

102 Ibid.

103 Ibid., 681.

104 Ibid., 682–83.

105 Ibid., 683–84.

106 Ibid., 684.

107 Ibid., 685.

108 Ibid., 268.

109 Newsom, 687.

110 Ibid., 649

111 Ibid., 643.

112 *Slaughter-House Cases*, 79.

113 Newsom, 679.

114 Ibid., 681

115 Ibid., 685.

116 Ibid.

117 Ross, *Justice of Shattered Dreams*, 201.

118 Ibid., 207.

119 Ibid., 207.

120 Ross, "Justice Miller's Reconstruction," 671.

121 Ross, *Justice of Shattered Dreams*, 207-8.

122 Ibid., 210.

123 Bryan H. Wildenthal, "How I Learned to Stop Worrying and Love the *Slaughter-House Cases*: An Essay in Constitutional-Historical Revisionism," *Thomas Jefferson Law Review* 23, no. 2 (Spring 2001), 244–45.

124 Ibid., 246

125 Newsom, 712.

126 Ibid.

127 Ibid., 721.

128 Ibid., 665.

129 Ross, "Justice Miller's Reconstruction," 676.

Book Review: Wendy Lower's *Hitler's Furies: German Women in the Nazi Killing Fields*

Gina Pittington

Hitler's Furies: German Women in the Nazi Killing Fields, by Wendy Lower. Boston: Mariner Books Houghton Mifflin Harcourt, 2014. ISBN 978-0-544-33449-6. Illustrations. Map. Notes. Index. Bibliography. Pp. xv, 280. Hardcover $26.00, Paperback $15.99, Electronic version available.

Wendy Lower is a Holocaust and genocide professor at Claremont McKenna College. Her research for *Hitler's Furies* began in 1992 when she visited Ukraine. The Soviet archives had recently opened for access, and Lower stationed herself near Heinrich Himmler's headquarters in Zhitomir. While there, she found multiple German records of the Nazi-occupied Eastern territory during World War II listing thousands of women that transferred to the eastern front beginning as early as 1941. While examining the newly released documents, she noticed women had served in multiple positions both civilian and conscripted. She compared them to the pioneers of the American West, as they were pioneers for Germany opening the new frontiers for extended living space.

Lower produces new research regarding women's roles on the eastern front and the brutality involved to carry out the Nazi ideology of living space and race purity, which included the extermination of European Jews. In *Hitler's Furies,* she explains the participation of women as accomplices and perpetrators in the Holocaust by examining thirteen separate women through thousands of wartime documents, interviews, and court data. Her focused questions on why these women killed and what happened to them after the war open new avenues for an investigation to trace further how female killers returned to normal life or disappeared entirely.

Complicit women that went to the East held positions as nurses, secretaries, teachers, and other positions outside of the camps, ghettos, and killing fields. Young German women, full of dreams and desire for independence from oppressive social traditions, volunteered for jobs with the possibility for advancement in the emerging Nazi empire abroad (9). Lower estimates that of the 640,000 women trained by the Red Cross, with the majority of the 400,000 trained for the wartime service, went East. Another 200,000 out of 500,000 moved east trained as radio operators, file-card keepers, flight recorders, and wire-tappers (6).

Women who killed came from different socioeconomic backgrounds, including wives of high-ranking officers, nurses giving lethal injections, secretaries signing death warrants, and prison guards at the killing centers. When gauging perpetrators, Lower contributes to prior studies, primarily Christopher Browning's 1992 work, titled *Ordinary Men, Reserve Battalion 101 and the Final Solution*. Much like Browning, Lower questions what drove them to kill and if they were programmed for murder, but from a feminist perspective. She argues that women married to SS officers and those employed as prison guards physically involved in killing Jews did not exhibit violent tendencies or sociopathic behavior before the war. They claimed to be a group unto themselves and their deeds are as grotesque as any that have been gathered into the history of the Holocaust (14).

The book contains fifty pages of notes outlining Lower's research and the in-depth study that went into the writing of *Hitler's Furies*, a scholarly work relating to women and gender social history, which encourages groundwork for further historical analysis. Lower provides a solid foundation of breaking the stereotype of German women serving under the Third Reich. Some research may be lacking, as many trails end up as cold cases: women did not face trials for war crimes or crimes against humanity, even as trials were held some three decades after the war. Most went on to lead normal lives, eventually disappearing with name changes due to marriage or moving from German territory following the war. *Hitler's Furies* provides a sound authority for women, gender, and World War II history.

Book Review: Kevin Gutzman's *Virginia's American Revolution: From Dominion to Republic, 1776–1840*

Matt Brent

Virginia's American Revolution: From Dominion to Republic, 1776-1840, by Kevin Gutzman. Lanham: Lexington Books, 2007. ISBN 978-0739121320. Pp. 248. Hardcover $104.00, Paperback $46.99.

In the late eighteenth century, the American British colonies developed a revolutionary fever caused by various conflicts and disagreements with the British crown. It was this excitement that drove the colonies to break with their mother country and seek their own destinies. One of those colonies, Virginia, the largest and most populous colony, is the center of Kevin Gutzman's *Virginia's American Revolution.* In his text, Gutzman explores the political and legal background to Virginia's desire for independence and seeks to explain how the attitude of Virginians affected the Commonwealth's development in the decade following independence.

While July 4, 1776 may be a date with which many Americans are familiar, Gutzman quickly points to a date two months prior. Although July may have been significant in that it resulted in the colonies collectively declaring independence from Britain, Virginia did so in May. The colony's declaration marked the Virginians' attempt to move towards self-government. Many established Virginians had argued that the colony had had the right to self-government since its founding, and that in establishing independence, it guaranteed that this right would continue to thrive. Accompanying this move toward political independence was the world's first written constitution.

Gutzman argues that the Virginians' attitude that they were Virginians first and only secondly members of a larger nation carries throughout the text and helps to explain the early actions of the state's leaders. It is refreshing, however, that Gutzman does not focus on prominent Virginians, like

George Washington, whom historians traditionally discuss. Rather, throughout his text, he places an emphasis on the actions of leaders within the state, such as Spencer Roane or John Taylor, who acted on behalf of the state.

Gutzman explains the desire for Virginia to maintain its ability to not only self-govern, but also to remain sovereign. As the United States moved forward and began to work towards a unification of states, there appeared to be hope that the nation would follow the Virginia way of government. Furthermore, the fact that many prominent Virginians held office at the national level led to the belief that the nation would, in fact, take Virginia's lead and model itself after its largest member. Many Virginians, argues Gutzman, supported the ratification of the national constitution because they believed it would allow Virginia to hold on to its sovereignty while establishing small compacts with its neighboring states. They believed Virginia would still be able to operate as it had over the previous few decades prior to the ratification of the Constitution.

Gutzman makes it clear that Virginians cared most about their own politics before those at the national level. The Old Dominion had established itself as a republican entity so that it could effectively handle the issues occurring within its borders, but what frightened the Virginia elite was the prospect of men of lower social standing gaining the opportunity to hold office. The upper classes of Virginia had traditionally held that the existence of different roles between the classes was essential to the success of the Commonwealth and maintained that the lower classes should show respect and reverence to those of higher class. With such great political change, social changes would inherently follow, and while he acknowledges some social change, Gutzman chooses to place his emphasis on the political and legal developments within his text.

The political discussions held within the text again surround the Virginian belief that Virginia was sovereign. Gutzman supports this argument with examples such as Jefferson's move to abolish established religion in the state. Jefferson claims, and Gutzman supports, the concept that had Virginia remained officially Anglican, Britain would still have had some sway within the state. By eliminating the official status of the Anglican Church, the state was able to maintain its sovereignty in this regard.

Gutzman also points to political conflicts between the Republicans and Federalists to demonstrate the Virginia attitude. When the federal government issued the Alien and Sedition Acts, Virginia, along with Kentucky, a state that had been carved out of Virginia, issued resolutions to proclaim the lack of constitutionality of those laws. These declarations effectively attempted to nullify the Acts and assert Virginia's long upheld sovereignty. Unfortunately for Virginia, as Gutzman's narrative describes, such sovereignty would later be assaulted through the courts with cases like *Martin v. Hunter's Lessee,* which allowed the federal courts to overturn rulings of state courts.

Another important note to address in Gutzman's work is his attention to the differences among Virginia's key players. He highlights the ideological disagreements between the states' rights advocate Patrick Henry and the staunch Federalist John Marshall, while also interjecting an examination of James Madison, the on-again, off-again Federalist. He discusses how upon Jefferson's election to the presidency, Virginians truly believed the nation would follow its lead. History has proven otherwise.

Gutzman contends that by examining Virginia separately from the other colonies, one can better understand its role in the American Revolution and understand its point of view. He effectively advocates his arguments by providing various examples and including the words of Virginia's Revolutionary leaders. While the Revolution offered benefits to Virginia, Gutzman explains that many of those leaders, upon reaching much older ages, questioned whether their actions truly benefited Virginia, as it moved away from an imperial crisis to an impending one. Although Gutzman does not provide an answer, he does adequately present their struggles through the course of his text.

Museum Review: Samegården Sami Museum, Kiruna, Sweden

Susan Danielsson
American Military University

The Samegården Sami Museum features exhibits on the little-known Sami, the Indigenous people of Northern Scandinavia.

Every year, thousands of tourists travel through Kiruna, Sweden to Abisko National Park to chase the Northern Lights or enjoy outdoor activities, such as dogsledding or wildlife safaris. Kiruna offers tourists an opportunity to learn about the little-known Sami, the Indigenous people of Northern Scandinavia, through the Samegården Sami Museum.

The Samegården Hotel operates this small museum in their basement. Despite its small size, its collection of artifacts and displays provide an excellent introduction to Sami culture and their struggles with the state from medieval to modern times. Upon entering the museum, visitors see a massive display of a traditional Sami home completely furnished with a reindeer skin bed, fire pit, and other household items. It provides an idea of how their home functioned in the arctic weather. To the right, there are displays describing ancient hunting practices, such as the pit system, where the Sami dug deep pits along the reindeers' natural walking paths. They covered the pits with moss and branches, encouraging reindeer to fall into the seemingly harmless area. The Sami used the pit systems throughout the Middle Ages.

Moving along the room, visitors discover artifacts related to the Sami pre-Christian religion, such as the shaman drum and reindeer bones. They had a polytheistic religion, believing that gods and spirits controlled natural forces, even illness and death. Reindeer herding and hunting had specific guardian spirits. Their exhibit explains that the *noaidi,* or Sami medicine man, played an important meditator role between the spiritual and natural world, with the shaman drum as his primary tool. It permitted him to enter a trance, where he interacted with the spirits. People called upon him in times of famine and disease to receive guidance from the spiritual realm. In the eighteenth century, Sweden banned the Sami from owning shaman drums to prevent them from practicing their old customs. Ansgar was the first Christian missionary to arrive in Sweden. In 829, he started to spread the faith at Birka, but Sweden did not fully convert to Christianity until the twelfth century. After conversion, the Swedish monarchy expected the Sami to convert too. The display recounts the story of Lars Nilsson, who ignored the state's warning against practicing the old religion. For his actions, the state condemned him to death. In 1693, his executioner threw his religious items into the fire, with Nilsson watching before he was burnt at the stake.

There are short information boards about the heavy tax burdens from the state and hunting competition with locals that helped to transform their lifestyle from nomads to settled peoples. Visitors are left with an understanding of early Sami culture and how they adapted to the political, economic, and social changes of Scandinavia. Other displays in this area provide basic information on village layouts and how they survived the harsh conditions of the Arctic.

After completing the displays to the right of the Sami home, visitors will want to backtrack and explore what is on the left side of the room. Once nomadic people, the Sami traveled compactly to more easily follow their reindeer herds on their yearly migration. There are numerous artifacts related to hunting and traveling, with each major artifact accompanied with a black and white image. For instance, there is a small boat neatly packed with essential items, with an accompanying photo of reindeer pulling boats across the snow. These photos put the items into perspective.

Visitors continue into a second room dedicated to the traditional clothing of the Sami. The clothes are laid out, allowing visitors to closely examine the elaborate textiles and colors of the pieces. Articles of clothing include hats, gloves, bags, and gender-specific pieces. One of the most interesting pieces is the brightly colored baby carrier that parents attached to the side of the reindeer to transport their children. Next to it, there are dolls with toy versions of the baby carriers. Unfortunately, there is little information presented in this room, but the museum offers some black and white photos for context. Photos show how the clothes were made and used in daily life, but it is mostly up to the visitors' imagination to draw conclusions about these pieces.

Many people outside of Scandinavia have never heard of the Sami, much less about their past mistreatment at the hands of the Swedish state. For its small size, the museum provides a wealth of information on this fascinating culture and shows that there is more to Scandinavia's history than just Vikings. Visitors can expect to spend up to an hour exploring all the displays, making it an excellent pit stop for people wanting to broaden their knowledge of the ancient people of Scandinavia.

Visiting the museum is free and open year-round. Displays are in English.

Sami Medicine Man, or Shaman photographed by Susan Danielsson

Lavvu, a Sami traditional home, photographed by Susan Danielsson

Sami traditional clothing, photographed by Susan Danielsson

Saber & Scroll Historical Journal

Submission Guidelines

To assist the SSHJ Editorial Staff in processing your manuscript efficiently, please follow these basic guidelines. Following these steps will ensure that our team can review your journal article promptly, and without additional work on your part. We reserve the right to reject, without further review, any submission that does not follow these guidelines or meet our high academic standards.

Thank you!
The SSHJ Editorial Staff

1. Manuscripts should be between 2,500 and 8,000 words in length, not including front or back matter, or between 800 and 1200 words for book reviews.

2. Only articles written using Microsoft Word (*.doc), and images in (*.jpeg) formats will be accepted; no *.PDF, *.TIFF or *.HEIC files.

3. The Journal's citation style is Chicago-Turabian, using end-notes, numbered in Arabic, super-script numerals, and including a complete, correctly formatted bibliography.

4. The narrative and notes are to be in 12-point Times New Roman, and the paragraphs double-spaced with a one-inch margin on all sides of the page.

5. A cover page, separate from the manuscript, must contain the following information; author's name, as they would like it to appear in print, school affiliation, if any, email address, and phone number.

6. The cover sheet must also include a statement affirming that the manuscript is not previously published material or submitted elsewhere for publication.

7. The author's name <u>should not appear</u> on the manuscript or book review to preserve the anonymity of the double-blind review process.

8. In addition to the narrative, each journal article must include a 100-150 word abstract and a list of 8-10 keywords.

9. SSHJ is always short of book reviews, so those received get immediate attention and are fast-tracked for publication.

1. We encourage authors to incorporate at least one, but no more than five, images into with their articles. Images might be photographs, paintings, illustrations, or maps

2. However, images must be of the author's creation or in the public domain. Smithsonian Open Access (https://www.si.edu/openaccess) is an excellent source.

3. A caption and source information must accompany every image. Ex: "Phineas Taylor Barnum, July 5, 1810–April 7, 1891, photographed by Matthew Brady. National Portrait Gallery NPG.97.63."

4. Our publisher prints the Journal using black text and grayscale images. Images should be approximately 300 dpi resolution.

5. SSHJ reserves the right to publish articles without any of the supplied images, substitute images that do not meet the minimum standards, or limit the number of images in an article or an issue because of space limitations.

To request additional information or submit complete articles for publication, email the Editor-in-Chief at EICSaberandScroll@apus.edu.

This publication is available open access at:
http://www.ipsonet.org/publications/open-access

Thanks to the generosity of the American Public University System

APU American Public University

AMU American Military University

DEMAND THE
IMPOSSIBLE

INTERNATIONAL OR
LOCAL OWNERSHIP?
DR. FLORIAN QEHAJA

DONALD J.
TRUMP'S
PRESIDENCY
International
Perspectives
Edited by
JOHN DIXON &
MAX J. SKIDMORE

Ongoing Issues in
Georgian Policy and
Public Administration

POVERTY
IN AMERICA

Made in the USA
Monee, IL
06 November 2026

67379283R10081